Solutions Manual

to accompany

Engineering Economy

Fifth Edition

Leland Blank, P. E.
Texas A&M University
and
American University of Sharjah, United Arab Emirates

Anthony Tarquin, P. E.
University of Texas at El Paso

Boston Burr Ridge, IL Dubuque, IA Madison, WI New York San Francisco St. Louis
Bangkok Bogotá Caracas Kuala Lumpur Lisbon London Madrid Mexico City
Milan Montreal New Delhi Santiago Seoul Singapore Sydney Taipei Toronto

McGraw-Hill Higher Education
A Division of The McGraw-Hill Companies

Solutions Manual to accompany
ENGINEERING ECONOMY, FIFTH EDITION
LELAND T. BLANK AND ANTHONY J. TARQUIN

Published by McGraw-Hill Higher Education, an imprint of The McGraw-Hill Companies, Inc., 1221 Avenue of the Americas, New York, NY 10020. Copyright © The McGraw-Hill Companies, Inc., 2002, 1998, 1989, 1983, 1976. All rights reserved.

The contents, or parts thereof, may be reproduced in print form solely for classroom use with ENGINEERING ECONOMY, provided such reproductions bear copyright notice, but may not be reproduced in any other form or for any other purpose without the prior written consent of The McGraw-Hill Companies, Inc., including, but not limited to, in any network or other electronic storage or transmission, or broadcast for distance learning.

This book is printed on acid-free paper.

2 3 4 5 6 7 8 9 0 QPD QPD 0 3 2 1

ISBN 0-07-243228-4

www.mhhe.com

Contents

Chapter 1-Foundations of Engineering Economy
Chapter 2-Factors: How Time and Interest Affect Money
Chapter 3-Combining Factors
Chapter 4-Nominal and Effective Interest Rates
Chapter 5-Present Worth Analysis
Chapter 6-Annual Worth Analysis
Chapter 7-Rate of Return Analysis: Single Alternative
Chapter 8-Rate of Return Analysis: Multiple Alternatives
Chapter 9-Benefit/Cost Analysis and Public Sector Economics
Chapter 10-Making Choices: The Method, MARR, and Multiple Attributes
Chapter 11-Replacement and Retention Decisions
Chapter 12-Selection from Independent Projects Under Budget Limitation
Chapter 13-Breakeven Analysis
Chapter 14-Effects of Inflation
Chapter 15-Cost Estimation and Indirect Cost Allocation
Chapter 16-Depreciation Methods
Chapter 17-After-Tax Economic Analysis
Chapter 18-Formalized Sensitivity Analysis and Expected Value Decisions
Chapter 19-More on Variation and Decision Making Under Risk
Appendix B-Basics of Accounting Reports and Business Ratios

INTRODUCTION

This document accompanies the text *Engineering Economy*, 5th edition by Leland Blank and Anthony Tarquin. It presents detailed solutions for the following end-of-chapter exercises:

- Problems
- FE (Fundamentals of Engineering) Review Problems
- Extended Exercises
- Case Studies

Appendix C of the textbook includes the final answers to approximately one-fourth of the problems only.

The website that supports the text also contains this document in a password protected format. Logon to the website at highered.mcgraw-hill.com/blank. It also contains updated information in engineering economics, spreadsheet exercises, additional FE review problems, self-test questions for each chapter of the text, and other materials. Additionally, sample lecture notes and self-help learning materials are included on the website for course instructors.

Reprint permission is hereby given to instructors to copy and distribute selected pages from this document only to individuals requiring solution to specific problems. These persons should be using the text in a course, workshop or seminar.

Report any significant errors or omissions in this document to either of the authors or McGraw-Hill so they may be corrected. The e-mail addresses for the authors are ATarquin@utep.edu and L-Blank@tamu.edu. Any additions and corrections will be posted on the website.

Chapter 1 – Foundations of Engineering Economy
Solutions to end of chapter exercises

Problems

1.1 A tangible factor is rather easily quantifiable while an intangible factor is not.

1.2 Three essential estimates in an engineering economic analysis are the amount of the cash flows involved, their time of occurrence, and the interest rate.

1.3 The seven steps are (1) Understand the problem (2) Collect information (3) Define alternatives (4) Identify the criteria (5) Evaluate the alternatives (6) Select the best alternative, and (7) Implement the solution.

1.4 Measures of worth are (1) Present worth (2) Annual worth (3) Future worth (4) Benefit/cost ratio (5) Rate of return, and (6) Capitalized cost.

1.5 Non-economic attributes that may be used in the decision-making process include social, environmental, legal, political, personal, etc.

1.6 The do-nothing alternative, if selected, would leave the situation as it presently is.

1.7 Interest rate = $\frac{2.25 - 2.0}{2.0} * 100$
= 12.5%

1.8 Profit = 6,000,000 (0.32)
= $1,920,000

1.9 1,600,000 = P + P (0.12)
1.12P = 1,600,000
P = $1,428,571

1.10 Earnings = 40,000,000 (0.25)
= $10,000,000

1.11 (a) Equivalent future amount = 10,000 + 10,000 (0.10)
= $11,000

(b) Equivalent past amount = $\frac{10,000}{1.10}$
= $9,090.91

1.12 Equivalent future cost = 14,500(1.12)
$$= \$16,240$$

The company should buy next year.

1.13 Move $40,000 two years ahead and set equal to $50,000:

$$40,000 (1 + i)(1 + i) = 50,000$$

$$(1 + i)^2 = \frac{50,000}{40,000}$$

$$i = 0.118$$
$$= 11.8\%$$

1.14 Assume a principal amount of $1,000:

Compound Amt. in 3 years = $1,000 (1 + 0.05)^3$
$$= \$1157.63$$

Simple Amount in 3 years = $1000 + 1000 (0.06)(3)$
$$= \$1180$$
Simple interest offer is better.

1.15 (a) $500,000 (i)(5) = 900,000$
$\quad\quad\quad i = 36\%$ per year simple interest

(b) $500,000 (1 + i)^5 = 900,000$
$\quad\quad i = 0.1247$
$\quad\quad = 12.47\%$ per year compound interest

1.16 $100,000 + 100,000 (0.10)(n) = 200,000$
$\quad\quad\quad n = 10$ years

1.17 (a) $P + P (0.10)(16) = 300,000$
$\quad\quad 2.6 P = 300,000$
$\quad\quad P = \$115,385$

(b) $P (1 + 0.10)^{16} = 300,000$
$\quad\quad P = \$65,289$

1.18 Under either arrangement, the company would owe $40,000 in interest each year and then pay back the $400,000 at the end of year 3. There would be *no difference* in the total amount repaid because there is no interest accumulated each year on the interest owed.

1.19 (a) Simple interest total amt = 1,750,000 (0.10)(5)
= $875,000

Compound interest total amt owed = 1,750,000 $(1 + 0.08)^4$
= $2,380,856

Total interest = 2,380,856 – 1,750,000
= $630,856

(b) The company should wait 1 year.

1.20 The symbols involved are as follows: P = P; F = 2P; n = 5, i = ?

1.21 The symbols are: P = ?, F = $50,000; n = 3; i = 15%

1.22 (a) FV(i%,n,A,P) finds the future value, F.
(b) IRR(first_cell:last_cell) finds the compound interest rate, i.
(c) PMT(i%,n,P,F) finds the equal periodic value, A.
(d) PV(i%,n,A,F) finds the present value, P.

1.23 The symbols and their values are as follows:

(a) F = ?; i = 8%; n = 10; A = $2,000; P = $10,000
(b) A = ?; i = 12%; n = 30; P = $16,000; F = 0
(c) P = ?; i = 9%, n = 15; A = $1,000; F = $700

1.24 The Excel symbols and their corresponding engineering economy symbols are as follows:

(a) FV = F (b) PMT = A (c) NPER = n (d) IRR = i
(e) PV = P

1.25 For built-in Excel functions, a parameter that does not apply can be left blank when it is not an interior one. For example, if there is no F involved when using the PMT function to solve a particular problem, it can be left blank because it is the end function. When the function that is not involved is an interior one (like the P in the PMT function), a comma must be put in its position.

1.26 Examples of safe investments are: savings accounts, certificates of deposit, government bonds, etc.
Examples of risky investments are: junk bonds, small start-up companies, and a relative's 'get-rich-quick' idea.

1.27 (a) Equity financing can come from cash on hand, stock sales, or retained earnings.

(b) Debt financing can come from bonds, loans, mortgages, venture capital, etc.

1.28 The lowest to highest yield is as follows: rate of return on safe investment, minimum attractive rate of return, cost of capital, rate of return on an investment.

1.29 Calculate the weighted average cost of capital, and set it equal to MARR. Select projects with a rate of return higher than this number.

WACC = (0.35)(16%) + (0.65)(9%) = 11.45%

MARR = 11.45%

Select the last two projects; 14% and 19% returns.

1.30 End of period convention means that the cash flows are assumed to have occurred at the end of the interest period in which they took place.

1.31 The difference between cash inflows and cash outflows is known as net cash flow.

1.32 (a) The borrower receives the money in January and April; positive cash flows. All others are negative.
   (b) The bank pays the money in January and April; negative cash flows. All others are positive.

| | Perspective | |
Month	Borrower	Bank
January	$1000	$-1000
February	-100	100
March	-100	100
April	400	-400
May	-150	150
June	-150	150
July	-1650	1650

1.33 The cash flow diagram is:

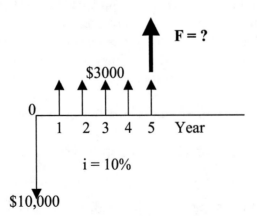

1.34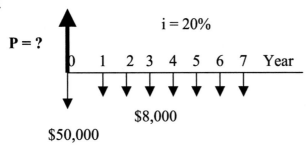

1.35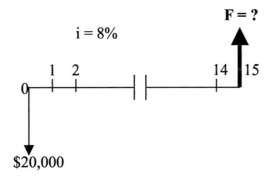

1.36 Time to double = 72/9
             = 8 years

1.37 Time to double = 72/8
             = 9 years

   Time to quadruple = (9)(2)
             = 18 years

1.38 5 = 72/i

   i = 72/5

    = 14.4% per year

1.39 The retirement account must quadruple in value before you can retire in 20 years. Therefore, it must double every 10 years. The estimated interest rate required is:

   i = 72/10

    = 7.2% per year

FE Review Solutions

1.40 Answer is (b)

1.41 Let P be the initial amount of money. Then, solve for F = 2P:

$$2P = P + P(0.10)(n)$$
$$P = P(0.10)(n)$$

$$n = 1/0.10$$
$$= 10 \text{ years}$$

Answer is (c)

1.42 $F = 10,000(1 + 0.10)^2$
       $= \$12,100$

Answer is (d)

1.43 $i = \dfrac{72}{6}$
       $= 12\%$

Answer is (d)

1.44 Answer is (c)

Extended Exercise Solution

1.

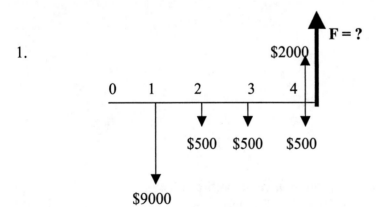

$F = [\{[-9000(1.08) - 500](1.08)\} - 500](1.08) + (2000-500)$
$= \$-10,960.60$

or $F = -9000(F/P,8\%,3) - 500(F/A,8\%,3) + 2000$

Chapter 1

2. A spreadsheet uses the FV function as shown in the formula bar. F = $–10,960.61.

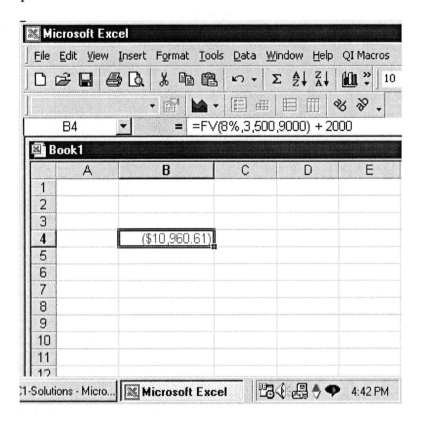

3. F = [{[–9000(1.08) – 300] (1.08)} – 500] (1.08) + (2000 –1000)
   = $–11,227.33

   Change is 2.02%. Largest maintenance charge is in the last year and, therefore, no compound interest is accumulated by it.

4. The fastest method is to use the spreadsheet function:

   FV(12.32%,3,500,9000) + 2000

   It displays the answer:

   F = $–12,445.43

Case Study Solution

There is no definitive answer to the case study exercises. The following are examples only.

1. The first four steps are: Define objective, information collection, alternative definition and estimates, and criteria for decision-making.

Objective: Select the most economic alternative that also meets requirements such as production rate, quality specifications, manufacturability for design specifications, etc.

Information: Each alternative must have estimates for life (likely 10 years), AOC and other costs (e.g., training), first cost, any salvage value, and the MARR. The debt versus equity capital question must be addressed, especially if more than $5 million is needed.

Alternatives: For both A and B, some of the required data to perform an analysis are:
- P and S must be estimated.
- AOC equal to about 8% of P must be verified.
- Training and other cost estimates (annual, periodic, one-time) must be finalized.
- Confirm n = 10 years for life of A and B.
- MARR will probably be in the 15% to 18% per year range.

Criteria: Can use either present worth or annual worth to select between A and B.

2. Consider these and others like them:
   - Debt capital availability and cost
   - Competition and size of market share required
   - Employee safety of plastics used in processing

3. With the addition of C, this is now a make/buy decision. Economic estimates needed are:
   - Cost of lease arrangement or unit cost, whatever is quoted.
   - Amount and length of time the arrangement is available.

   Some non-economic factors may be:
   - Guarantee of available time as needed.
   - Compatibility with current equipment and designs.
   - Readiness of the company to enter the market now versus later.

Chapter 2 – Factors: How Time and Interest Affect Money
Solutions to end of chapter exercises

Problems

2.1 (1) From Table 15:
(P/F,10%,28) = 0.0693

(2) From Table 8:
(A/P,3%,6) = 0.18460

(3) From Table 14:
(F/A,9%,20) = 51.1601

(4) From Table 22:
(F/P,20%,28) = 164.8447

(5) From Table 26:
(P/A,30%,15) = 3.2682

2.2 F = 150,000(F/P,12%,3)
= 150,000(1.4049)
= $210,735

2.3 F = 1,700,000(F/P,20%,2)
= 1,700,000(1.4400)
= $2,448,000

2.4 F = 25,000(F/P,15% 4)
= 25,000(1.7490)
= $43,725

2.5 P = 65,000(P/F,12%,5)
= 65,000(0.5674)
= $36,881

2.6 P = 75(P/F,15%,3)
= 75(0.6575)
= $49.31 million

2.7 P = 65,000(P/F,10%,3)
= 65,000(0.7513)
= $48,835

2.8   F = 162,000(F/P,8%,6)
      = 162,000(1.5869)
      = $257,078

2.9   F = 150,000(F/P,7%,9)
      = 150,000(1.8385)
      = $275,775

2.10 P = 6000(P/F,10%,2) + 9000(P/F,10%,3) + 5000(P/F,10%,6)
      = 6000(0.8264) + 9000(0.7513) + 5000(0.5645)
      = $14,543

2.11 P = 50,000(P/F,8%,2) + 125,000(P/F,8%,5)
      = 0.5(0.8573) + 1.25(0.6806)
      = $127,940

2.12 F = 25,000(F/P,5%,6)
      = 25,000(1.3401)
      = $33,503

2.13 A = 1.75 million (A/P,12%,10)
      = 1.75 million (0.17698)
      = $309,715

2.14 P = 280,000(P/A,20%,10)
      = 280,000(4.1925)
      = $1,173,900

2.15 P = 140,000(P/A,10%,5)
      = 140,000(3.7908)
      = $530,712

2.16 P = 55,000(P/A,15%,5)
      = 55,000(3.3522)
      = $184,371

2.17 Savings per year = (458 − 360)(0.90)(20,000)
                       = $1,764,000

      P = 1,764,000(P/A,8%,5)
        = 1,764,000(3.9927)
        = $7,043,123

2.18 F = 225,000(F/A,12%,3)
      = 225,000(3.3744)
      = $759,240

2.19 F = 2000(F/A,8%,35)
     = 2000(172.3168)
     = $344,634

2.20 A = 250,000(A/F,10%,4)
     = 250,000(0.21547)
     = $53,868

2.21 F = 200,000(F/A,10%,5)
     = 200,000(6.1051)
     = $1,221,020

2.22 (1a) Interpolate between n = 32 and n = 34 in Table 21.

$$32 = 199.6293$$
$$33 = x$$
$$34 = 277.9638$$

$$\frac{33-32}{34-32} = \frac{c}{277.9638 - 199.6293}$$

$$2c = 78.3345$$
$$c = 39.16725$$

(F/P,18%,33) = 199.6293 + 39.1673
            = 238.7966

(1b) (F/P,18%,33) = $(1 + 0.18)^{33}$
                   = 235.5625

(2a) Interpolate between n = 35 and n = 40 in Table 9

$$35 = 0.05358$$
$$37 = x$$
$$40 = 0.05052$$

$$\frac{2}{5} = \frac{c}{0.05358 - 0.05052}$$

c = 0.00122

(A/P,4%,7) = 0.05358 − 0.00122
          = 0.05236

Chapter 2

(2b) $(A/P, 4\%, 37) = \dfrac{0.04(1+0.04)^{37}}{(1+0.04)^{37}-1}$
$= 0.05224$

(3a) Interpolate between n = 50 and n = 55.

$50 = 94.8889$
$54 = x$
$55 = 96.5619$

$\dfrac{4}{5} = \dfrac{c}{96.5619 - 94.8889}$

$c = 1.3384$

$(P/G, 10\%, 54) = 94.8889 + 1.3384$
$= 96.2273$

(3b) $(P/G, 10\%, 54) = \dfrac{(1+0.10)^{54} - 0.10(54) - 1}{(0.10)^2 (1+0.10)^{54}}$
$= 96.2763$

2.23 (1a) Interpolate between i = 16% and i = 18%.

$16 = 19.4608$
$17 = x$
$18 = 27.3930$

$\dfrac{1}{2} = \dfrac{c}{27.3930 - 19.4608}$

$c = 3.9661$

$(F/P, 17\%, 20) = 19.4608 + 3.9661$
$= 23.4269$

(1b) $(F/P, 17\%, 20) = (1 + 0.17)^{20}$
$= 23.1056$

(2a) Interpolate between i = 20% and i = 22%.

$20 = 3.8372$
$21 = x$
$22 = 3.6193$

Chapter 2

$$\frac{1}{2} = \frac{c}{3.8372 - 3.6193}$$

c = 0.1090

(P/A,21%,8) = 3.8372 − 0.1090
= 3.7282

(2b) $(P/A,21\%,8) = \frac{(1 + 0.21)^8 - 1}{0.21(1+0.21)^8}$

= 3.7256

(3a) Interpolate between i = 25% and i = 30%.

25 = 3.6698
28 = x
30 = 3.1718

$$\frac{3}{5} = \frac{c}{3.6698 - 3.1718}$$

c = 0.2988

(A/G,28%,18) = 3.6698 − 0.2988
= 3.3710

(3b) $(A/G,28\%,18) = \left[\frac{1}{0.28} - \frac{18}{(1 + 0.28)^{18} - 1} \right]$

= 3.3573

2.24 (a) G = $1000   (b) Cash flow is indicated by CF
    (c) n = 10 $CF_8 = 3000 + (8-1)(1000) = \$10,000$

2.25 (a) G = $350   (b) $CF_6 = 50 + 350(6) = \$2150$
    (c) n = 12

2.26 (a) G = $−100   (b) $CF_5 = 8000 - 100(5) = \$7500$

2.27 30,000 = 12,000 + G(A/G,10%,5)
    18,000 = G(1.8101)
        G = $9,944 per year

2.28 (a) $CF_n = 180,000 - (n-1)G$

CF in year 7 is 0; therefore

$0 = 180,000 - (7-1)G$
$6G = 180,000$
$G = \$30,000$ per year

(b) $P = 180,000(P/A,12\%,6) - 30,000(P/G,12\%,6)$
$= 180,000(4.1114) - 30,000(8.9302)$
$= \$472,146$

2.29 (a) $CF_3 = 4000 + 800(3-1)$
$= \$5600$

(b) $A = 4000 + 800(A/G,10\%,5)$
$= 4000 + 800(1.8101)$
$= \$5448$ per year

2.30 (a) $P = 12,000(P/A,15\%,10) + 1000(P/G,15\%,10)$
$= 12,000(5.0188) + 1000(16.9795)$
$= \$77,205$

(b) $A = 12,000 + 1000(A/G,15\%,10)$
$= 12,000 + 1000(3.3832)$
$= \$15,383$ per year

2.31 $P = 4000(P/A,12\%,4) + 850(P/G,12\%,4)$
$= 4000(3.0373) + 850(4.1273)$
$= \$15,657$

2.32 (a) Cost in prior year = $2000/(1-0.80)$
$= \$10,000$

(b) $P = 2000(P/A,18\%,5) + 230(P/G,18\%,5)$
$= 2000(3.1272) + 230(5.2312)$
$= \$7458$

2.33 Move F to year zero and write PW equation for cash flow.

$6000(P/F,15\%,4) = 200(P/A,15\%,4) + G(P/G,15\%,4)$
$6000(0.5718) = 200(2.8550) + G(3.7864)$
$G = \$755$

2.34 Convert $50 million into A using A/F and then set equal to annual worth of cash flow.

$$50,000,000(A/F,12\%,6) = 6,000,000 + G(A/G,12\%,6)$$
$$50,000,000(0.12323) = 6,000,000 + G(2.1720)$$
$$G = \$74,355$$

2.35 $P = 4,000,000(P/A,15\%,) + 500,000(P/G,15\%,5) - [1,000,000(P/A,15\%,5) - 100,000(P/G,15\%,5)]$
   $= 3,000,000(3.3522) + 600,000(5.7751)$
   $= \$13,521,660$

2.36 Bonus money year 1 $= 5,000,000(0.01)$
   $= \$50,000$

$$P = 50,000 \left[1 - \frac{(1+0.20)^5}{(1+0.10)^5} \right] \Big/ (0.10 - 0.20)$$

   $= \$272,525$

2.37 Deposit in year 1 $= 50,000(0.10)$
   $= \$5000$

   Find P and then convert to F in year 15

$$P = 5000 \left[1 - \frac{(1+0.04)^{15}}{(1+0.12)^{15}} \right] \Big/ (0.12 - 0.04)$$

   $= \$41,936$

   $F = 41,936(F/P,12\%,15)$
   $= 41,936(5.4736)$
   $= \$229,541$

2.38 Value of stock in year 1 $= 500(30)$
   $= \$15,000$

$$P = 15,000 \left[1 - \frac{(1+0.15)^5}{(1+0.10)^5} \right] \Big/ (0.10 - 0.15)$$
   $= \$74,668$

2.39 $\text{Factor}_1 = \left[1 - \frac{(1+0.06)^1}{(1+0.10)^1} \right] \Big/ (0.10 - 0.06) \, (F/P,10\%,1)$

   $= 1.0000$

$$\text{Factor}_2 = \left[1 - \frac{(1+0.06)^2}{(1+0.10)^2}\right] / (0.10 - 0.06) \, (F/P, 10\%, 2)$$

$$= 2.1600$$

$$\text{Factor}_3 = \left[1 - \frac{(1+0.06)^3}{(1+0.10)^3}\right] / (0.10 - 0.06) \, (F/P, 10\%, 3)$$

$$= 3.4996$$

2.40 First find P and then convert to F in year 7:

$$P = 2000 \left[1 - \frac{(1+0.10)^7}{(1+0.15)^7}\right] / (0.15 - 0.10)$$

$$= \$10,696$$

$$F = 10,696 (F/P, 15\%, 7)$$
$$= 10,696 (2.6600)$$
$$= \$28,451$$

2.41 Cash flow in year 1 is A_1.

$$65,000 = A_1 \left[1 - \frac{(1+0.09)^{10}}{(1+0.15)^{10}}\right] / (0.15 - 0.09)$$

$$65,000 = A_1 [1 - 0.58518] / 0.06$$
$$65,000 = 6.9137 A_1$$
$$A_1 = \$9402$$

2.42 First find P and then solve for CF in year 1, which is identified by A_1.

$$P = 38,000 \, (P/F, 12\%, 5)$$
$$= 38,000 \, (0.5674)$$
$$= \$21,561$$

$$21,561 = A_1 \left[1 - \frac{(1+0.05)^5}{(1+0.12)^5}\right] / (0.12 - 0.05)$$

$$21,561 = 3.9401 A_1$$
$$A_1 = \$5472$$

2.43 $P = \dfrac{1000(20)}{1 + 0.10}$
       $= \$18,182$

   $F = 18,182(F/P,10\%,20)$
       $= 18,182(6.7275)$
       $= \$122,319$

2.44 $P = 2000\left[1 - \dfrac{(1+0.05)^4}{(1+0.08)^4}\right] / (0.08 - 0.05)$

   $= \$7104.44$

   $F = 7104(F/P,8\%,4)$
       $= 7104(1.3605)$
       $= \$9666$

2.45 Each year the cash flow (CF) grows by 1.05. The CF in years 3 and 2 are:

   CF in year 3 $= \dfrac{1250}{1.05} = \$1190.48$

   CF in year 2 $= \dfrac{1190.48}{1.05} = \$1133.79$

2.46 (a) $150,000 = 5000(F/A,i,10)$
       $(F/A,i,0) = 30.0$

   From tables, i is between 22% and 24%. Interpolate or use the indicated spreadsheet function.

   i = 22.9% per year   (IRR function)

   (b) $150,000 = 2(5000)(F/A,i,10)$
       $(F/A,i,10) = 15.0$

   i is between 8% and 9%
   i = 8.7%   (RATE function)

2.47 $10,000(F/P,i,4) = 15,000$
       $(F/P,i,4) = 1.5000$

   From tables, i is between 10% and 11%. Interpolate, use formula, or use a spreadsheet function

   i = 10.7% per year   (RATE function)

Chapter 2

2.48 $72,000 = 30,000(P/A,i,3)$
       $(P/A,i,3) = 2.4000$

   Interpolate between i = 12% and 14% or use a spreadsheet function.
   i = 12.04% per year   (IRR function)

2.49 $150,000(F/P,i,6) = (250,000 - 30,000)$
            $(F/P,i,6) = 1.4667$

   Interpolate between 6% and 7% or use a spreadsheet function.
   i = 6.59% per year   (RATE function)

2.50 Set-up PW equation.

   $85,000 = [(50,000(P/A,i,5) + 40,000(P/G,i,5)] - [50,000(P/A,i,5) + 10,000(P/G,i,5)]$

   $85,000 = 30,000(P/G,i,5)$
   $(P/G,i,5) = 2.8333$

   Interpolate between 35% and 40% or use a spreadsheet function.

   i = 39.05% per year   (IRR function)

2.51 $170,000 = 30,000(P/A,12\%,n)$
       $(P/A,12\%,n) = 5.6667$

   From Table 17, n is between 10 and 11 years.

   n = 10.05 years (or 11 as an integer value)   (NPER function)

2.52 $1,000,000 = 100,000(P/A,8\%,n)$
      $(P/A,8\%,n) = 10.000$

   From Table 13, n is between 20 and 21 years.

   n = 20.9 years (or 21 years)   (NPER function)

2.53 $1,000,000 = 12,000(F/A,15\%,n)$
       $(F/A,15\%,n) = 83.3333$
       From Table 19, n is between 18 and 19 years. (NPER function)

   n = 18.6 years   (NPER function)

   This is 8.6 years from now since she started 10 years ago.

2.54 $135{,}000 = 25{,}000 \left[1 - \dfrac{(1+0.25)^n}{(1+0.20)^n} \right] \Big/ (0.20 - 0.25)$

$\left[\dfrac{1.25}{1.20} \right]^n = 1.27$

$[1.0417]^n = 1.27$

$n \log 1.0417 = \log 1.27$
$n (0.017728) = 0.1038$
$\quad\quad n = 5.8$ (or 6 years)

Spreadsheet: IRR function, i = 20.92% after 6 years.

2.55 $10A = A \ (F/A, 8\%, n)$
$(F/A, 8\%, n) = 10.000$

From Table 13, n is between 7 and 8 years.

$n = 7.6$ or 8 years   (NPER function)

FE Review Solutions

2.56 $F = 10{,}000(F/P, 10\%, 20)$
$\quad\quad = 10{,}000(6.7275)$
$\quad\quad = \$67{,}275$

Answer is (d)

2.57 $P = 20{,}000(P/F, 12\%, 10)$
$\quad\quad = 20{,}000(0.3220)$
$\quad\quad = \$6440$

Answer is (a)

2.58 $P = 50{,}000(P/F, 18\%, 7)$
$\quad\quad = 50{,}000(0.3139)$
$\quad\quad = \$15{,}695$

Answer is (a)

2.59 A = 100,000(A/P,12%,10)
      = 100,000(0.17698)
      = $17,698

Answer is (c)

2.60 P = 100,000(P/A,18%,5)
      = 100,000(3.1272)
      = $312,720
Answer is (b)

2.61 A = 100,000(A/F,10%,5)
      = 100,000(0.16380)
      = $16,380

Answer is (c)

2.62 F = 6000(F/A,8%,30)
      = 6000(113.2832)
      = $679,699

Answer is (d)

2.63 P = 5000(P/F,12%,3) + 10,000(P/F,12%,5) + 10,000(P/F,12%,8)
      = 5000(0.7118) + 10,000(0.5674) + 10,000(0.4039)
      = $13,272

Answer is (b)

2.64 P = 5000(P/A,10%,5) + 1000(P/G,10%,5)
      = 5,000(3.7908) + 1000(6.8618)
      = $25,816

Answer is (d)

2.65 F = [8000(P/A,10%,10) + 500(P/G,10%,10)](F/P,10%,10)
      = [8000(6.1446) + 500(22.8913)](2.5937)
      = $157,185

Answer is (c)

2.66 P = 100,000(P/A,10%,5) − 10,000(P/G,10%,5)
      = 100,000(3.7908) − 10,000(6.8618)
      = $310,462

Answer is (a)

2.67 100,000 = 25,000(P/A,i,10)
     (P/A,i,10) = 4.0000

From interest tables, i is between 20% and 22%

Answer is (c)

2.68 50,000 = 10,000(P/A,18%,n)
     (P/A,18%,n) = 5.0000

From Table 21, n is between 13 and 14 years

n = 14 years

Answer is (d)

Case Study Solution

I. Manhattan Island

Simple interest

n = 375 years from 1626 – 2001

$$P + I = P + nPi = 375(24)(.06) + 24$$
$$= P(1 + ni) = 24(1 + 375(.06))$$
$$= \$564$$

Compound interest

$$F = P(F/P,6\%,375)$$
$$= 24(3,088,157,729.0)$$
$$= \$74,115,785,490 \Rightarrow \$74+ \text{ billion}$$

II. Stock-option plan

1. Years 0 1 ... 5 ... 35

   $50/mth = 60 deposits

   F = ? after 5 years

   F = ? after 35 years

   Age 22 27 57

2. Value when leaving the company

$$F = A(F/A, 1.25\%, 60)$$
$$= 50(88.5745)$$
$$= \$4428.73$$

3. Value at age 57 (n = 30 years)

$$F = P(F/P, 15\%, 30)$$
$$= 4428.73(66.2118)$$
$$= \$293,234$$

4. Amount for 7 years to accumulate F = $293,234

$$A = F(A/F, 15\%, 7)$$
$$= 293,234(.09036)$$
$$= \$26,497 \text{ per year}$$

5. Amount in 20's: 5(12)50 = $3000
Amount in 50's: 7(26,497) = $185,479

Chapter 3 – Combining Factors
Solutions for end of chapter exercises

Problems

3.1 $P = (95)(2000)(P/A,20\%,3) + 125(2000)(P/A,20\%,6)(P/F,20\%,3)$
  $= 190,000(2.1065) + 250,000(3.3255)(0.5787)$
  $= \$881,352$

3.2 $P = 200,000(P/A,15\%,2) + 300,000(P/A,15\%,3)(P/F,15\%,2)$
  $= 200,000(1.6257) + 300,000(2.2832)(0.7561)$
  $= \$843,038$

3.3 $P = 15,000(P/A,10\%,5)(P/F,10\%,1)$
  $= 15,000(3.7908)(0.9091)$
  $= \$51,693$

3.4 $P_{17} = 20,000(P/A,8\%,4)$
  $= 20,000(3.3121)$
  $= \$66,242$

  $A = 66,242(A/F,8\%,18)$
  $= 66,242(0.02670)$
  $= \$1769$

3.5 $P_{-1} = (500)(3,000)(P/F,8\%,1)$
  $= 1,500,000(0.9259)$
  $= \$1,388,850$

  $A = 1,388,850(A/P,8\%,3)$
  $= 1,388,850(0.38803)$
  $= \$538,915$

3.6 $F_{30} = 6000(F/A,8\%,31)$
  $= 6000(123.3459)$
  $= \$740,075$

  $P_{29} = 740,075(P/F,8\%,1)$
  $= 740,075(0.9259)$
  $= \$685,236$

  $A = 685,236(A/P,8\%,30)$
  $= 685,236(0.08883)$
  $= \$60,869$

3.7 (a) P = 13,000 (P/A,6%,10) (P/F,6%,2)
= 13,000 (7.3601) (0.8900)
= $85,156

(b) A = 85,156(P/A,6%,12)
= 85,156(0.11928)
= $10,157

3.8 P = 80,000 (P/A,10%,5) (P/F,10%,1)
= 80,000 (3.7908) (0.9091)
= $275,697

A = 275,697(A/P,10%,6)
= 275,697(0.22961)
= $63,303

3.9 Annual savings = 1,000,000 (0.15)
= $150,000

Engineer's earnings = 150,000 (0.75)
= 112,500 per year

F_7 = 112,500 (F/A,10%,7)
= 112,500 (9.4872)
= 1,067,310

A = 1,067,310 (A/F,10%,5)
= 1,067,310 (0.16380)
= $174,825

3.10 P = 28,000 (P/A,8%,3) + 48,000 (P/A,8%,7) (P/F,8%,3)
= 28,000 (2.5771) + 48,000 (5.2064) (0.7938)
= $270,535

A = 270,535(A/P,8%,10)
= 270,535(0.14903)
= $40,318

3.11 P = 600 (P/A,12%,2) + 4000 (P/A,12%,4) (P/F,12%,2)
= 600 (1.6901) + 4000 (3.0373) (0.7972)
= $10,699

A = 10,699(A/P,12%,6)
= 10,699(0.24323)
= $2602

3.12 $P_{10} = 66,000(P/F,8\%,8)$
      $= 66,000(0.5403)$
      $= \$35,660$

   $A = 35,660(A/F,8\%,8)$
   $= 35,660(0.09401)$
   $= \$3352$

3.13 $F = 10,000\ (F/P,8\%,14) + 600\ (F/A,8\%,6)\ (F/P,8\%,8) + 700\ (F/A,8\%,5)\ (F/P,8\%,3)$
   $= 10,000\ (2.9372) + 600\ (7.3359)\ (1.8509) + 700\ (5.8666)\ (1.2597)$
   $= \$42,692$

3.14 $P = -9000 - 1000\ (P/A,18\%,5) + 3000\ (P/A,18\%,9)\ (P/F,18\%,5)$
   $= -9000 - 1000\ (3.1272) + 3000\ (4.3030)\ (0.4371)$
   $= \$-6485$

3.15 Let F_5 be the amount in year 5

   $F_5 = -500\ (F/A,12\%,5)\ (F/P,12\%,1) - 800\ (P/A,12\%,10)$
   $= -500\ (6.3528)\ (1.1200) - 800\ (5.6502)$
   $= \$-8078$

3.16 $P = -10,000 - 17,000(P/F,12\%,3) - 21,000(P/F,12\%,5) + 15,000(P/A,12\%,5) + 17,000(P/A,12\%,5)\ (P/F,12\%,5)$
   $= -10,000 - 17,000\ (0.7118) - 21,000\ (0.5674) + 15,000\ (3.6048) + 17,000\ (3.6048)\ (0.5674)$
   $= \$54,827$

3.17 Deposit $= 10,000(P/A,12\%,7)\ (P/F,12\%,12)$
   $= 10,000\ (4.5638)\ (0.2567)$
   $= \$11,715$

3.18 Size of fund now $= 20,000\ (F/A,15\%,5)$
   $= 20,000\ (6.7424)$
   $= \$134,848$

   Let F_9 be the future worth of \$134,848 in year 9
   $F_9 = 134,848\ (F/P,15\%,4)$
   $= 134,848\ (1.7490)$
   $= \$235,849$

   The \$50,000 deposit must grow to $350,000 - 235,849 = \$114,151$. This will take n years.

$114,151 = 50,000 \, (F/P,15\%,n)$
$(F/P,15\%, n) = 2.28302$

From 15% interest table, n is between 5 and 6 years, and close to 6 (actually 5.9 years). Therefore, deposit must be made 2 years from now.

3.19 $P_0 = 800 - 500 \, (P/F,14\%,2) + 1000 \, (P/A,14\%,2)(P/F,14\%,2) - x \, (P/F,14\%,5)$
$\quad\quad + 1200 \, (P/A,14\%,2) \, (P/F,14\%,5) - 3x \, (P/F,14\%,8)$

$= 800 - 500 \, (0.7695) + 1000 \, (1.6467) \, (0.7695) - x \, (0.5194) + 1200 \, (1.6467) \, (0.5194) - 3x \, (0.3506)$

$1.5712 \, x = 2708.74$
$\quad\quad x = \$1724 \text{ (cost)}$

3.20 $P = -20,000 \, (P/A,12\%,3) - 25,000 \, (P/A,12\%,3) \, (P/F,12\%,3) - 30,000 \, (P/A,12\%,3) \, (P/F,12\%,6)$
$\quad = \$-127,279$

$A = -127,279 \, (A/P,12\%,9)$
$\quad = -127,279 \, (0.18768)$
$\quad = \$-23,888$

3.21 $P = -1200 \, (P/A,12\%,3) \, (P/F,12\%,1) - 2000 \, (P/A,12\%,5) \, (P/F,12\%,4) - 1000 \, (P/F,12\%,8)$
$\quad = -1200 \, (2.4018) \, (0.8929) - 2000 \, (3.6048) \, (0.6355) - 1000 \, (0.4039)$
$\quad = \$-7559$

$A = -7,559 \, (A/P,12\%,9)$
$\quad = -7,559 \, (0.18768)$
$\quad = \$-1419$

3.22 $40,000 = 4000 \, (P/A,15\%,7) \, (P/F,15\%,2) + x \, (P/F,15\%,11)$
$40,000 = 4000 \, (4.1604) \, (0.7561) + x \, (0.2149)$
$\quad\quad x = \$127,582$

3.23 Let amount in year 8 be F_8.
$F_8 = 900 \, (F/A,16\%,4) \, (F/P,16\%,5) + 1300 \, (F/A,16\%,2) \, (F/P,16\%,3)$
$\quad - 1300 \, (F/P,16\%,2) + 500 \, (F/P,16\%,1) + 900 + 900 \, (P/F,16\%,1)$
$= 900 \, (5.0665) \, (2.1003) + 1300 \, (2.1600) \, (1.5609) - 1300 \, (1.3456)$
$\quad + 500 \, (1.1600) + 900 + 900 \, (0.8621)$
$= \$14,467$

3.24 P = –5000 – 3500 (P/A,12%,3) – 5000 (P/A,12%,6) (P/F,12%,3) – 15,000 (P/F,12%,10)
= –5000 – 3500 (2.4018) – 5000 (4.1114) (0.7118) – 15,000 (0.3220)
= $–32,869

A = –32,869(A/P,12%,10)
= –32,869(0.17698)
= $–5817

3.25 Move cash flow to year 8 and set equal to $8000:

8000 = 2000 (F/A,15%,5) (F/P,15%,4) + 4000 (F/P,15%,6) + x (F/A,15%,4) + 500 (P/A,15%,3) + 400 (P/F,15%,1)

8000 = 2000 (6.7424) (1.7490) + 4000 (2.3131) + x (4.9934) + 500 (2.2832) + 400 (0.8696)

4.9934 x = –26,327
x = $–5272

3.26 P = (35) (100,000) (P/F,12%,1) + [(35) (100,000) (P/A,12%,9) – 100,000 (P/G,12%,9)] (P/F,12%,1)
= 3,500,000 (0.8929) + [3,500,000 (5.3282) – 100,000 (17.3563)] (0.8929)
= $18,226,830

3.27 Savings is PW cost of accidents plus crossing gates.

Cost of accidents = 250,000 (P/A,12%,10)
= 250,000 (5.6502)
= $1,412,550

Cost of gates = 15,000 (P/A,12%,10) + 1000 (P/G,12%,10)
= 15,000 (5.6502) + 1000 (20.2541)
= $105,007

Savings = $1,517,557

Cost to build = $1,200,000

The overpass should be built.

3.28 P = –19,000 – 15,000 (P/A,15%, 3) – 10,000 (P/F,15%,3) – [6000 (P/A,15%,4) + 1000 (P/G,15%,4)] (P/F,15%,3)
= –19,000 – 15,000 (2.2832) –10,000 (0.6575) –[6000(2.8550) + 1000 (3.7864)] (0.6575)
= $–73,576

$$A = -73{,}576 \, (A/P,15\%,7)$$
$$= -73{,}576 \, (0.24036)$$
$$= \$-17{,}685$$

3.29 $P = -10{,}000 + 4000 \, (P/F,12\%,1) - 2000 \, (P/F,12\%,2) + [4000 \, (P/A,12\%,6)$
$\quad + 1000 \, (P/G,12\%,6)] \, (P/F,12\%,2) - 8000 \, (P/F,12\%,6)$

$\quad = -10{,}000 + 4000 \, (0.8929) - 2000 \, (0.7972) + [4000 \, (4.1114) + 1000 \, (8.9302)]$
$\quad \quad (0.7972) - 8000 \, (0.5066)$

$\quad = \$8154$

$A = 8154 \, (A/P,12\%,8)$
$\quad = 8154 \, (0.20130)$
$\quad = \$1641 \quad \quad \quad (\$1{,}641{,}000)$

3.30 $P = 4000 + 4000 \, (P/F,18\%,1) + [4000 \, (P/A,18\%,6) + 1000 \, (P/G,18\%,6)]$
$\quad (P/F,18\%,1) + [9100 \, (P/A,18\%,4) + 100 \, (P/G,18\%,4)] \, (P/F,18\%,7)$

$\quad = 4000 + 4000 \, (0.8475) + [4000 \, (3.4976) + 1000 \, (7.0834)] \, (0.8475)$
$\quad \quad + [9100 \, (2.6901) + 100 \, (3.4828)] \, (0.3139)$

$\quad = \$33{,}044$

3.31 Let x = size of first payment

$40{,}000 = x \, (P/A,10\%,2) + (x + 2000) \, (P/F,10\%,3) + (x + 3000) \, (P/F,10\%,4)$
$\quad \quad \quad + (x + 4000) \, (P/F,10\%,5)$

$40{,}000 = x \, (1.7355) + (x + 2000) \, (0.7513) + (x + 3000) \, (0.6830) + (x + 4000)$
$\quad \quad \quad (0.6209)$

$3.7907 \, x = 33{,}965$
$\quad x = \$8960$

3.32 Set future worth of annual \$2000 amounts in year 3 equal to present value of cash flows:

$2000 \, (P/A,12\%,9) \, (F/P,12\%,3) = [200 \, (P/A,12\%,3) + 100 \, (P/G,12\%,3)]$
$\quad \quad \quad \quad \quad \quad \quad \quad \quad \quad \quad \quad (F/P,12\%,4) + x + 500 \, (P/A,12\%,6) + 100$
$\quad \quad \quad \quad \quad \quad \quad \quad \quad \quad \quad \quad (P/G,12\%,6)$

$2000 \, (5.3282) \, (1.4049) = [200 \, (2.4018) + 100 \, (2.2208)] \, (1.5735)$
$\quad \quad \quad \quad \quad \quad \quad \quad \quad \quad + x + 500 \, (4.1114) + 100 \, (8.9302)$

$14{,}971 = 4054 + x$
$\quad x = \$10{,}917$

3.33 Tuition cost in years 0 through 3 = $4000 (5) = $20,000 per year

$P = 20,000 + 20,000(P/A,8\%,2) +$
$\quad \{20,000[1 - (1.10/1.08)^{18}]/(0.08 - 0.10)\}(P/F,8\%,17)$
$= 20,000 + 20,000 (1.7833) + 20,000 (19.5682)(0.2703)$
$= \$161,452$

3.34 $P_{-1} = 20,000[(1 - (1.06/1.14)^{11})/(0.14 - 0.06)]$
$= \$137,707$

$P = 137,707(F/P, 14\%, 1)$
$= 137,707(1.14)$
$= \$156,986$

$A = 156,986(A/P,14\%,10)$
$= 156,986(0.19171)$
$= \$30,096$

3.35 First find present worth of annual cost in year 4 and then move back to year 0 with other costs:
$P_3 = -13,000[(1 - (1.10/1.20)^7)/(0.20 - 0.10)]$
$= \$-59,299$

$P = -29,000 - 13,000(P/A,20\%,3) - 59,299 (P/F,20\%,3)$
$= -29,000 - 13,000(2.1065) - 59,299 (0.5787)$
$= \$-90,701$

3.36 $P_{-1} = -15,000[(1 - (1.10/1.16)^5)/(0.16 - 1.10)]$
$= \$-58,304$

$A = -58,304 (A/P,16\%,5)$
$= -58,304 (0.30541)$
$= \$-17,807$

3.37 $P = -73,000 - 21,000[(1 - (1.05/1.18)^9)/(0.18 - 0.05)]$
$= -73,000 - 105,039$
$= \$-178,039$

3.38 $P = -340 (P/F,12\%,2) - [500 (P/A,12\%,9) - 20 (P/G,12\%,9)] (P/F,12\%,2)$
$= -340 (0.7972) - [500 (5.3282) - 20 (17.3563)] (0.7972)$
$= \$-2118$

3.39 $P = 5000 + 1000 (P/A,15\%,4) + [900 (P/A,15\%,6) - 100 (P/G,15\%,6)] (P/F,15\%,4)$
$= 5000 + 1000 (2.8550) + [900 (3.7845) - 100 (7.9368)] (0.5718)$
$= \$9349$

$$A = 9{,}349 \ (A/P, 15\%, 10)$$
$$= 9{,}349 \ (0.19925)$$
$$= \$1863$$

3.40 $P = 2000 + 2000 \ (P/A, 12\%, 6) + 500 \ (P/F, 12\%, 5) + [3500 \ (P/A, 12\%, 6) - 200 \ (P/G, 12\%, 6)] \ (P/F, 12\%, 6)$

$= 2000 + 2000 \ (4.1114) + 500(0.5674) + [3500(4.1114) - 200 \ (8.9302)](0.5066)$
$= \$16{,}892$

$A = 16{,}892 \ (A/P, 12\%, 12)$
$= 16{,}892 \ (0.16144)$
$= \$2727$

3.41 $20{,}000 = 5000 \ (P/A, 8\%, n) - 500 \ (P/G, 8\%, n)$

Solve by trial and error

Try n = 10:	$20,000 < $20,562	n too large
n = 9:	$20,000 < $20,330	n too large
n = 8:	$20,000 > $19,830	n too small

The account will be depleted in the 9th year.

3.42 $P = -500 - [\ 480 \ (P/A, 15\%, 8) - 20 \ (P/G, 5\%, 8)]$
$= -500 - [\ 480 \ (4.4873) - 20 \ (12.4807)]$
$= \$-2404$

3.43 $P_1 = - [1200 \ (P/A, 12\%, 3) + 200 \ (P/G, 12\%, 3)] - [1300 \ (P/A, 12\%, 5) - 500 \ (P/G, 12\%, 5)] \ (P/F, 12\%, 4)$

$= -1200 \ (2.4018) - 200 \ (2.2208) - [1300 \ (3.6048) - 500 \ (6.3970)] \ (0.6355)$
$= \$-4272$

$F = -4272 \ (F/P, 12\%, 9)$
$= -4272 \ (2.7731)$
$= \$-11{,}846$

FE Review Solutions

3.44 $F = 2000 \ (F/P, 10\%, 10)$
$= 2000 \ (2.5937)$
$= \$5187$

Answer is (b)

3.45 F = 10,000 (F/P,10%,5) − 4,000
      = 10,000 (1.6105) − 4000
      = $12,105

Answer is (a)

3.46 P = 1,000 (P/A,10%,25) (P/F,10%,4)
      = 1,000 (9.0770) (0.6830)
      = $6200

Answer is (a)

3.47 P = 1000 (P/A,10%,5) + 2000 (P/A,10%,5) (P/F,10%,5)
      = 1000 (3.7908) + 2000 (3.7908) (0.6209)
      = $8498

Answer is (c)

3.48 A = 100,000 (P/F,10%,10) (A/P,10%,5)
      = 100,000 (0.3855) (0.26380)
      = $10,169

Answer is (d)

3.49 P_2 = 1000 (P/A,10%,8) + 500 (P/G,10%,8)
      = 1000 (5.3349) + 500 (16.0287)
      = $13,349

F_{10} = 13,349 (F/P,10%,8)
      = 13,349 (2.1436)
      = $28,615

Answer is (c)

3.50 F = 5000 (F/P,10%,10) + 7000 (F/P,10%,7) + 2000 (F/A,10%,7)
      = 5000 (2.5937) + 7000 (1.9487) + 2000 (9.4872)
      = $45,584
Answer is (b)

Extended Exercise Solution

Solution by Hand

Cash flows for purchases at g = −25% start in year 0 at $4 million. Cash flows for parks development at G = $100,000 start in year 4 at $550,000. All cash flow signs in the solution are +.

| | Cash flow | |
Year	Land	Parks
0	$4,000,000	
1	3,000,000	
2	2,250,000	
3	1,678,000	
4	1,265.625	$550,000
5	949,219	650,000
6		750,000

1. Find P for all project funds (in $ million)

   P = 4 + 3(P/F,7%,1) + ... + 0.750(P/F,7%,6)
     = 13.1716   ($13,171,600)

   Amount to raise in years 1 and 2:

   A = (13.1716 − 3.0)(A/P,7%,2)
     = (10.1716)(0.55309)
     = 5.6258   ($5,625,800 per year)

2. Find remaining project fund needs in year 3, then find the A for the next 3 years (years 4, 5, and 6):

   F_3 = (13.1716 − 3.0)(F/P,7%,3)
      = (10.1716)(1.2250)
      = 12.46019

   A = 12.46019(A/P,7%,3)
     = 12.46019(0.38105)
     = 4.748   ($4,748,000 per year)

Solution by computer

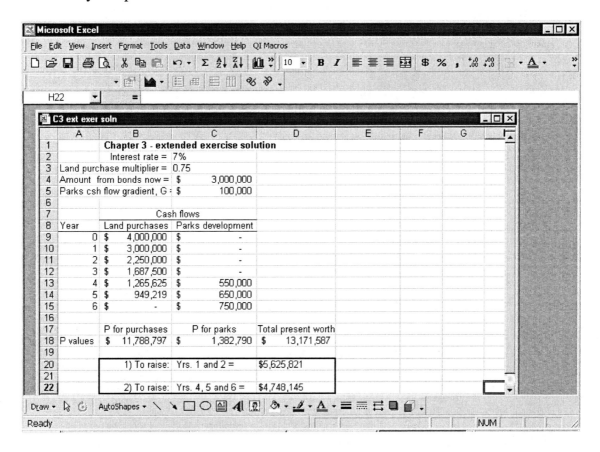

Chapter 4 – Nominal and Effective Interest Rates
Solutions to end of chapter exercises

Problems

4.1 (a)   r/quarter = 0.50 * 3
             = 1.5% per quarter

   (b)   r/year = 0.50 * 12
            = 6% per year

   (c)   r/2 years = 0.50 * 24
                = 12% per 2 years

4.2 (a)   i/month   = 18/12
                   = 1.5%
        r/2 months = 1.5 * 2
                   = 3% per 2 months

   (b)   r/6 months = 1.5 * 6
                  = 9% per 6 months

   (c)   r/2 years   = 1.5 * 24
                    = 36% per 2 years

Note all rates are compounded monthly.

4.3 (a)   i/quarter   = 12/4
                   = 3% per quarter

        r/6 months = 3 * 2
                   = 6% per 6 months

   (b)   r/2 years   = 3 * 8
                    = 24% per 2 years

4.4 (a) effective   (b) nominal   (c) effective
   (d) effective   (e) nominal

4.5   i/year = $(1 + 0.16/4)^4 - 1$
           = 16.99% per year

4.6   i/year = $(1 + 0.13/12)^{12} - 1$
           = 13.80% per year

4.7 Use Equation [4.5] or [4.8] with i = r/12 and find r where i_a = effective i = 0.12

$0.12 = (1 + r/12)^{12} - 1$
i = r / 12 = 0.009489
r = 11.387% per year

4.8 $16.986\% = (1 + 0.16/m)^m - 1$

Solve for m by trial and error

try m = 2:   16.986 ≠ 16.64
try m = 4:   16.986 = 16.986

Therefore, compounding period is quarterly

4.9 $i_a = (1 + 0.08/4)^4 - 1$
   = 8.24% per year

F = 1000(F/P,8.24%,10)
  = $1000(1 + 0.0824)^{10}$
  = 1000(2.2074)
  = $2207.40

4.10 r/2 months = 1 * 2
              = 2% per 2 months

Use Equation [4.8]
effective i = $(1 + 0.02/2)^2 - 1$
            = 2.01% per 2 months

4.11 i/month = 18/12
            = 1.5%

r/quarter = 1.5 * 3
         = 4.5% per quarter

effective i = $(1 + 0.045/3)^3 - 1$
            = 4.568% per quarter

4.12 effective i = 0.06707
                = $(1 + r/26)^{26} - 1$

r/6 months = 6.5%
   i/week = 6.5%/26
         = 0.25% per week
Weekly rate is an effective rate.

4.13 Payment period is monthly; compounding period is quarterly.

4.14 Payment period is monthly; compounding period is daily.

4.15 $i_a = (1 + 0.12/4)^4 - 1$
  $= 12.55\%$ per year

  $(P/F, 12.55\%, 4) = \dfrac{1}{(1 + 0.1255)^4}$
  $= 0.6232$

4.16 Five years is 60 months. Therefore,

  $P = 2.8(P/F, 4\%, 60)$
   $= 2.8(0.0951)$
   $= \$0.26628$ billion
   $= \$266.28$ million

4.17 $i = 10\%$ per 6 months

  $P = 1.2(P/F, 10\%, 1)$
   $= 1.2(0.9091)$
   $= \$1.0909$ million

4.18 Use P/F or F/P factor with n = 120 and solve for "i". Using F/P:

  $6.9 = 3.9 (F/P, i, 120)$
  $(F/P, i, 120) = 1.7692$

  From interest tables, i is between 0.25% and 0.50%

  Solve from equation for F/P:
  $1.7692 = (1 + i)^{120}$
    $i = 0.477\%$ per month

4.19 PP = CP = month
   i/month = 24/12
      = 2%

  Monthly savings = 1000 (250-150)
          = $100,000 per month

  $P = 100,000(P/A, 2\%, 24)$
   $= 100,000(18.9139)$
   $= \$1,891,390$

4.20 i/month = 18/12
             = 1.5%

   P = 50(20)(P/F,1.5%,3)
     = 1000(0.9563)
     = $956.3 million

4.21 i/quarter = 20/4
               = 5%

   P = 2.3(P/F,5%,12)
     = 2.3(0.5568)
     = $1.28 million

4.22 i/6 months = 16/2
                = 8%

   F = 10,000(F/P,8%,6) + 25,000(F/P,8%,4) + 30,000
     = 10,000(1.5869) + 25,000(1.3605) + 30,000
     = $79,881
   or
   effective i = $(1 + 0.16/2)^2 - 1 = 16.64\%$

   F = 10,000(F/P,16.64%,3) + 25,000(F/P,16.64%,2) + 30,000
     = $79,881

4.23 Cost/month = 2.99 * 2
                 = $5.98

   P = 5.98(P/A,1½%,10)
     = 5.98(9.2222)
     = $55.15

4.24 A = 300(A/P,1%,12)
        = 300(0.08885)
        = $26.655 million per month

4.25 A = 20,000(A/P,3%,8)
        = 20,000(0.14246)
        = $2849 per quarter

4.26 The subsidy is equal to the foregone interest.

   F = 100,000(F/A,0.5%,8)(F/P,0.5%,2)
     = 100,000(8.1414)(1.0100)
     = $822,281

Actual cost = (100,000)(8)
= $800,000

Subsidy = 822,281 – 800,000
= $22,281

4.27 P = 14.99(P/A,1%,24)
= 14.99(21.2434)
= $318.44

4.28 PP > CP; find effective i per PP = year.
i/year = $(1 + 0.15/2)^2 - 1$
= 15.56%

P = 150,000[1– $(1.20/1.1556)^5$ /(0.1556–0.20)]
= 150,000[4.67226]
= $700,839

4.29 PP > CP; find effective i per PP = 6 months.

i/6 months = $(1 + 0.06/6)^6 - 1$
= 6.15%

P = 80(P/A,6.15%,8) + 2(P/G,6.15%,8)
= 80(6.1731) + 2(19.6790)
= $533.21   (factors obtained from equations)

Can use NPV function at 6.15% once series $80 through $94 is entered.

4.30 PP > CP; find effective i per PP = quarter.

i/quarter = $(1 + 0.03/3)^3 - 1$
= 3.03%

A = 2 (A/P, 3.03%, 20)
= 2 (0.0674)
= $0.134805 million
= $134,805 per quarter

4.31 PP and CP are monthly.
F = 1000(F/P,2%,36) + 2000(P/A,2%,12)(F/P,2%,36)
    + 3000(P/A,2%,16)(F/P,2%,24)
= 1000(2.0399) + 2000(10.5753)(2.0399)
    + 3000(13.5777)(1.6084)
= $110,700

4.32 PP > CP; find effective i per PP = quarter.

$$i/\text{quarter} = (1 + 0.18/12)^3 - 1$$
$$= 4.57\%$$

$$P = 1000(P/F,4.57\%,1) + 2000(P/A,4.57\%,2)(P/F,4.57\%,1)$$
$$+ 3000(P/A,4.57\%,4)(P/F,4.57\%,4)$$

$$= 1000(0.9563) + 2000(1.8708)(0.9563)$$
$$+ 3000(3.5817)(0.8363)$$
$$= \$13,521$$

$$A = 13,521(A/P,4.57\%,8)$$
$$= 13,521(0.15204)$$
$$= \$2056 \text{ per quarter}$$

The spreadsheet function, after entering cash flows into A1 through A9, is PMT(4.57%,8,–NPV(4.57%,A2:A9) + A1).

4.33 A/quarter = (300)(3)
     = $900

$$F = 900(F/A,1.5\%,60)$$
$$= 900(96.2147)$$
$$= \$86,593$$

4.34 Move $1000 withdrawal in month 2 to time 0. Move $1,000 withdrawal in month 11 to month 9. Move $1000 withdrawal in month 23 to month 21.

$$F = (10,000 - 1000)(F/P,2\%,8) - 1000(F/P,2\%,5) - 1000(F/P,2\%,1)$$
$$= 9000(1.1717) - 1000(1.1041) - 1000(1.0200)$$
$$= \$8421$$

4.35 Move withdrawals to beginning of each year and deposits to end of each year. Deposits at end of year 1 are $2600 and withdrawal is $2600. At end of year 2, deposit is $1000 and withdrawal is $1000. Therefore, $F_3 = 0$.

4.36 $0.132 = e^r - 1$
     $e^r = 1.132$
     r = 12.399% per year

4.37 r = .02/month = 0.06 / quarter

effective $i = e^{0.06} - 1$
     = 6.18% per quarter

4.38 $0.036 = e^r - 1$
  $e^r = 1.036$
  $r = 3.537\%$ per quarter

  r/month = 3.537/3
  = 1.18% per month

4.39 $i/year = e^{0.12} - 1$
  = 12.75%

  P = 10,000(P/A,12.75%,6)
  = 10,000(4.0255)
  = $40,255

4.40 $i/year = e^{0.10} - 1$
  = 10.517% per year

  F = 20,000(F/P,10.517%,10)
  = 20,000(2.7183)
  = $54,365

4.41 $i/quarter = e^{0.03} - 1$
  = 3.045%

  F = 75,000(A/F,3.045%,12)
  = 75,000(0.07028)
  = $5271

4.42 $i/month = e^{0.015} - 1$
  = 1.511% per month

  P = 500(P/A,1.511%,36) + 10(P/G,1.511%,36)
  = 500(27.6101) + 10(438.7074)
  = $18,192

4.43 F = 10,000(F/P,12%,3)(F/P,15%,2)
  = 10,000(1.4049)(1.3225)
  = $18,580

4.44 First find P in year 4 and then move back to year 0.

  P = 50,000(P/F,1%,36)(P/F,10%,4)
  = 50,000(0.6989(0.6830)
  =$23,867

4.45 P = 5000 + 6000(P/A,12%,4) + 9000(P/F,20%,1)(P/F,12%,4)
   = 5000 + 6000(3.0373) + 9000(0.8333)(0.6355)
   = $27,990

4.46 P = 10,000 + 5000(P/A,10%,3) + 7000(P/A,12%,2)(P/F,10%,3)
   = 10,000 + 5000 (2.4869) + 7000(1.6901)(0.7513)
   = $31,323

Now, substitute A values for all previous F values:

31,323 = 10,000 + A(2.4869) + A(1.6901)(0.7513)
21,323 = 3.7567 A
   A = $5,676

FE Review Solutions

4.47 (c)

4.48 (d)

4.49 (a)

4.50 (b)

4.51 (b)

4.52 $i = (1 + 0.15/12)^{12} - 1$
   $= 16.075\%$

Answer is (c)

4.53 $i = e^{0.18} - 1$
   $= 19.72\%$

Answer is (b)

4.54 P = 40,000(P/F,5%,8)
   = 40,000(0.6768)
   = $27,072

Answer is (a)

4.55 A = 800,000(A/P,1%,36)
   = 800,000(0.03321)
   = $26,568
Answer is (d)

4.56 A = 500,000(A/F,9%,12)
     = 500,000(0.04965)
     = $24,825

Answer is (b)

Case Study Solution

1. **Plan C: 15-Year Rate** The calculations for this plan are the same as those for plan A, except that i = 9 ½% per year and n = 180 periods instead of 360. However, for a 5% down payment, the P&I is now $1488.04 which will yield a total payment of $1788.04. This is greater than the $1600 maximum payment available. Therefore, the down payment will have to be increased to $25,500, making the loan amount $124,500. This will make the P&I amount $1300.06 for a total monthly payment of $1600.06.

   The amount of money required up front is now $28,245 (the origination fee has also changed). The plan C values for F_{1C}, F_{2C}, and F_{3C} are shown below.

   $$F_{1C} = (40,000 - 28,245)(F/P,0.25\%,120)$$
   $$= \$15,861.65$$

   $$F_{2C} = 0$$

   $$F_{3C} = 170,000 - [124,500(F/P,9.5\%/12,120) - 1300.06(F/A,9.5\%/12,120)]$$
   $$= \$108,097.93$$

   $$F_C = F_{1C} + F_{2C} + F_{3C}$$
   $$= \$123,959.58$$

   The future worth of Plan C is considerably higher than either Plan A ($87,233) or Plan B ($91,674). Therefore, Plan C with a 15-year fixed rate is the preferred financing method.

2. <u>Plan A</u>

   Loan amount = $142,500
   Balance after 10 years = $129,582.48
   Equity = 142,500 – 129,582.48 = $12,917.52

   Total payment made = 1250.56(120) = $150,067.20

   Interest paid = 150,067.20 – 12,917.52 = $137,149.68

3.   Amount paid through first 3 yrs = 1146.58 (36) = $41,276.88
Principal reduction through first 3 yrs = 142,500 – 139,297.08 = $3,202.92
Interest paid first 3 yrs = 41,276.88 – 3202.92 = $38,073.96

Amount paid year 4 = 1195.67(12) = 14,348.04
Principal reduction year 4 = 139,297.08 – 138,132.42 = 1164.66
Interest paid year 4 = 14,348.04 – 1164.66 = 13,183.38

Total interest paid in 4 years = 38,073.96 + 13,183.38 = $51,257.34

4. Let DP = down payment
Fixed fees = 300 + 200 + 200 + 350 + 150 + 300 = $1500
Available for DP = 40,000 – 1500 – (loan amount)(0.01)
where loan amount = 150,000 – DP

DP = 40,000 – 1500 - [(150,000 – DP)(0.01)]
     = 40,000 – 1500 – 1500 + 0.01DP
0.99DP = 37,000
DP = $37,373.73

check: origination fee = (150,000 – 37,373.73)(0.01) = 1126.26
available DP = 40,000 – 1500 – 1126.26 = $37,373.73

5. Amount financed = $142,500

Monthly P&I @ 10% = $1,250.56

Monthly P&I @ 11% = 142,500(A/P,11%/12, 60)

$$A = (142,500)\left[\frac{(0.009167)(1 + 0.009167)^{360}}{(1 + 0.009167)^{360} - 1}\right] = \$1357.06$$

Monthly P&I @ 12% = $1465.77

Monthly P&I @ 13% = $1576.33

Monthly P&I @ 14% = $1688.44

Increase from one interest rate to the other:

106.50
108.71
110.56
112.11

Increase varies: 10% to 11% = $106.50
                             11% to 12% =   108.71
                             12% to 13% =   110.56
                             13% to 14% =   112.11

6. In buying down interest, you must give lender money now instead of money later. Therefore, to go from 10% to 9%, lender must recover the additional 1% now.

P&I @ 10% = 1250.54
P&I @ 9% = 1146.59

Difference = $103.95/month

P = 103.95(P/A,10%/12,360)
  = 103.95(113.9508)
  = $11,845.19

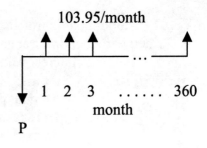

Chapter 5 – Present Worth Analysis
Solutions to end of chapter exercises

Problems

5.1 When projects are mutually exclusive, only one will be chosen. When projects are independent, more than one may be chosen.

5.2 The do-nothing alternative is usually an option when the alternatives are revenue alternatives.

5.3 For revenue projects, the revenues are dependent upon the project. Revenues are not assumed equal between projects, as they are for service projects.

5.4 (a) Revenue (b) Revenue (c) Service (d) Revenue (e) Service

5.5 (a) m = 5; maximum number of alternatives is $2^5 = 32$.

(b) No 3, 4 or 5 at a time alternatives are possible. Only 12 are acceptable:

DN, 1, 2, 3, 4, 5, 13, 14, 15, 23, 24, and 25.

5.6 PW_{US} = -22,000 – 2000(P/A,15%,3) + 12,000(P/F,15%,3)
  = -22,000 – 2000(2.2832) + 12,000(0.6575)
  = $-18,676

PW_J = -26,000 – 1200(P/A,15%,3) + 15,000(P/F,15%,3)
  = -26,000 – 1200(2.2832) + 15,000(0.6575)
  = $-18,877

Purchase the US model.

5.7 PW_{solar} = -$12,600 – 1,400(P/A,14%,4)
  = -$12,600 – 1,400(2.9137)
  = $-16,679

PW_{line} = -11,000 – 800(P/A,14%,4)
  = -11,000 – 800(2.9137)
  = $-13,331

Install the power line.

5.8 PW_A = -300,000 – 900,000(P/A,12%,10)
  = -300,000 – 900,000(5.6502)
  = $-5,385,180

$PW_B = -1,200,000 - (200,000 + 150,000)(P/A,12\%,10)$
   $= -1,200,000 - 350,000(5.6502)$
   $= \$-3,177,570$

The company should select Plan B.

5.9 $PW_A = -2,000,000 - 5000(P/A,1\frac{1}{2}\%,36) + 200,000(P/F,1\frac{1}{2}\%,36)$
   $= -2,000,000 - 5000(27.6607) + 200,000(0.5851)$
   $= \$-2,021,280$

$PW_B = -22,000 - 45,000(P/A,1\frac{1}{2}\%,6) - 10,000(P/A,1\frac{1}{2}\%,30)(P/F,1\frac{1}{2}\%,6)$
   $= -22,000 - 45,000(5.6972) - 10,000(24.0158)(0.9145)$
   $= \$-497,998$

Select Plan B.

5.10 $PW_{small} = -1,700,000 - 12,000(P/A,1\%,120) + 170,000(P/F,1\%,120)$
   $= -1,700,000 - 12,000(69.7005) + 170,000(0.3030)$
   $= \$-2,484,896$

$PW_{large} = -2,100,000 - 9,000(P/A,1\%,120) + 210,000(P/F,1\%,120)$
   $= -2,100,000 - 9,000(69.7005) + 210,000(0.3030)$
   $= \$-2,663,675$

Select the small pipe size.

5.11 The LCM is 6 years.

$PW_A = -25,000 - 25,000\,[(P/F,15\%,2) + (P/F,15\%,4)] - 4000(P/A,15\%,6)$
   $= -25,000 - 25,000\,[0.7561 + 0.5718] - 4000(3.7845)$
   $= \$-77,336$

$PW_B = -88,000 - 1,400(P/A,15\%,6)$
   $= -88,000 - 1,400(3.7845)$
   $= \$-93,298$

Select Plan A.

5.12 The LCM is 4 years.

$PW_X = -7,650 - 7,650(P/F,16\%,2) - 1,200(P/A,16\%,4)$
   $= -7,650 - 7,650(0.7432) - 1,200(2.7982)$
   $= \$-16,693$

$PW_Y = -12{,}900 - 900(P/A,16\%,4) + 2000(P/F,16\%,4)$
$= -12{,}900 - 900(2.7982) + 2000(0.5523)$
$= \$-14{,}314$

Select type Y.

5.13 $PW_{JX} = -15{,}000 - 15{,}000(P/F,18\%,3) - 9000(P/A,18\%,6) + 2000(P/F,18\%,3)$
$\quad\quad + 2000(P/F, 18\%, 6)$
$= -15{,}000 - 15{,}000(0.6086) - 9{,}000(3.4976) + 2000(0.6083) + 2000(0.3704)$
$= \$-53{,}649$

$PW_{KZ} = -35{,}000 - 7{,}000(P/A,18\%,6) + 20{,}000(P/F,18\%,6)$
$= -35{,}000 - 7{,}000(3.4976) + 20{,}000(0.3704)$
$= \$-52{,}075$

Select material KZ.

5.14 The least common multiple of lives is 6. Therefore, use 6 year comparison period.

$PW_{land} = -110{,}000 - 110{,}000(P/F,15\%,3) - 95{,}000(P/A,15\%,6)$
$\quad\quad + 15{,}000[(P/F,15\%,3) + (P/F,15\%,6)]$
$= -110{,}000 - 110{,}000(0.6575) - 95{,}000(3.7845) + 15{,}000(0.6575 + 0.4323)$
$= \$-525{,}506$

$PW_{incinerate} = -800{,}000 - 60{,}000(P/A,15\%,6) + 250{,}000(P/F,15\%,6)$
$= -800{,}000 - 60{,}000(3.7845) + 250{,}000(0.4323)$
$= \$-918{,}995$

$PW_{contract} = -190{,}000(P/A,15\%,6)$
$= -190{,}000(3.7845)$
$= \$-719{,}055$

Select land application.

5.15 $FW_x = -82{,}000(F/P,15\%,3) - 30{,}000(F/A,15\%,3) + 40{,}000$
$= -82{,}000(1.5209) - 30{,}000(3.4725) + 40{,}000$
$= \$-188{,}889$

$FW_y = -97{,}000(F/P,15\%,3) - 27{,}000(F/A,15\%,3) + 50{,}000$
$= -97{,}000(1.5209) - 27{,}000(3.4725) + 50{,}000$
$= \$191{,}285$

Select Robot X.

5.16 Calculate FW's over least common multiple of 6 years:

FW_C = -40,000(F/P,20%,6) - 40,000(F/P,20%,3) - 10,000(F/A,20%,6)
  + 12,000(F/P,20%,3) + 12,000
 = -40,000(2.9860) - 40,000(1.7280) - 10,000(9.9299) + 12,000(1.7280)
  + 12,000
 = $-255,123

FW_D = -65,000(F/P,20%,6) - 12,000(F/A,20%,6) + 25,000
 = -65,000(2.9860) - 12,000(9.9299) + 25,000
 = $-288,294

Select machine C.

5.17 FW_K = -160,000(F/P,1%,48) - 160,000(F/P,1%,24) - 7000(F/A,1%,48)
  + 40,000(F/P,1%,24) + 40,000
 = -160,000(1.6122) - 160,000(1.2697) - 7000(61.2226) + 40,000(1.2697)
  + 40,000
 = $-798,874

FW_L = -210,000(F/P,1%,48) - 5000(F/A,1%,48) + 26,000
 = -210,000(1.6122) - 5000(61.2226) + 26,000
 = $-618,675

Select Process L.

5.18 FW_Q = -42,000(F/P,15%,4) - 42,000(F/P,15%,2) - 6000(F/A,15%,4)
 = -42,000(1.7490) - 42,000(1.3225) - 6,000(4.9934)
 = $-158,963

FW_R = [-80,000 - 7000(P/A,15%,4) - 1000(P/G,15%,4) + 4000(P/F,15%,4)]
  (F/P,15%,4)
 = [-80,000 - 7000(2.8550) - 1000(3.7864) + 4000(0.5718)](1.7490)
 = $-177,496

Select Project Q.

5.19 PW_{cap} = -100,000 - 25,000(P/A,10%,5) - $\dfrac{50,000}{0.10}$(P/F,10%,5)

 = -100,000 - 25,000(3.7908) - 500,000(0.6209)
 = $-505,220

5.20 Find AW through one life cycle, then divide by i:

$$AW = -32,000(A/P,14\%,4) - 6000 + 8000(A/F,14\%,4)$$
$$= -32,000(0.34320) - 6000 + 8000(0.20320)$$
$$= \$-15,356.80$$

$$CC = PW_{cap} = -15,356.80/0.14$$
$$= \$-109,691$$

5.21 Find the P value (P_3) at the end of year 3 sufficient to produce $100,000 per year. Then find P in year zero.

$$P_3 = 5(20,000)/0.08$$
$$= \$1,250,000$$

$$P_0 = 1,250,000(P/F,8\%,3)$$
$$= 1,250,000(0.7938)$$
$$= \$992,250$$

5.22 $PW_{100} = 10,000(P/A,10\%,100)$
    $= 10,000(9.9993)$
    $= \$99,993$

$$CC = PW_{cap} = 10,000/0.10$$
$$= \$100,000$$

Difference = 100,000 – 99,993
           = $7

5.23 $P_{pipe} = -200 - 6/0.10$
    $= \$-260$ million

$P_{canal} = -325 - 1/0.10$
    $= \$-335$ million

Construct the pipeline conduit.

5.24 First find AW for alternative V and then divide by i.

$$AW_V = -50,000(A/P,10\%,10) - 30,000 + 10,000(A/F,10\%,10)$$
$$= -50,000(0.16275) - 30,000 + 10,000(0.06275)$$
$$= \$-37,510$$

$$CC_V = -37,510/0.10$$
$$= \$-375,100$$

For alternative W, CC of the salvage is zero, since n is ∞.

$CC_W = -500,000 - 1000/0.10$
$\quad\quad = \$-510,000$

Select Alternative V.

5.25 First find A for alternatives E and F and then divide by i.

$AW_E = -50,000(A/P,6\%,4) - 30,000 + 5000(A/F,6\%,4)$
$\quad\quad = -50,000(0.28859) - 30,000 + 5000(0.22859)$
$\quad\quad = \$-43,287$

$CC_E = -43,287/0.06$
$\quad\quad = \$-721,450$

$AW_F = -300,000(A/P,6\%,8) - 10,000 + 70,000(A/F,6\%,8)$
$\quad\quad = -300,000(0.16104) - 10,000 + 70,000(0.10104)$
$\quad\quad = \$-51,239$

$CC_F = -51,239/0.06$
$\quad\quad = \$-853,983$

The CC of the salvage value for G is zero, since n is ∞.
$CC_G = -900,000 - 3,000/0.06$
$\quad\quad = \$-950,000$

Select Alternative E.

5.26 Determine the future principal amount F_{11} accumulated in year 11.

$F_{11} = 10,000(F/P,15\%,11) + 30,000(F/P,15\%,8) + 8000(F/A,15\%,5)(F/P,15\%,3)$
$\quad\quad = 10,000(4.6524) + 30,000(3.0590) + 8000(6.7424)(1.5209)$
$\quad\quad = \$220,330$

Re-label F_{11} to P relative to the perpetual withdrawal series, then multiply by i.

$A = P\,i = 220,330(0.15)$
$\quad\quad = \$33,050$ per year forever

5.27 (a) $n_p = 200,000/(90,000 - 50,000)$
$\quad\quad\quad = 200,000/40,000$
$\quad\quad\quad = 5$ years

(b) $0 = -200{,}000 + (90{,}000 - 50{,}000)(P/A,15\%,n)$
$(P/A,15\%,n) = 5.0000$

From the 15% interest tables, n_p is between 9 and 10 years.
$n_p = 9.98$ years, therefore 10 years.

5.28 (a) Semi-Automatic:
$n = 40{,}000/10{,}000$
$= 4$ years

Automatic:
$n = 90{,}000/15{,}000$
$= 6$ years

Retain Semi-Automatic machine.

(b) Semi-Automatic
$0 = -40{,}000 + 10{,}000(P/A,10\%,n)$
$(P/A,10\%,n) = 4.0000$
From 10% table, n is between 5 and 6 years
$n_p = 5.4$ years

Automatic:
$0 = -90{,}000 + 15{,}000 (P/A,10\%,n)$
$(P/A,10\%,n) = 6.0000$
From 10% table, n is between 9 and 10 years
$n_p = 9.6$ years

Retain neither alternative.

5.29 Payback analysis:
Alternative A: $0 = -600{,}000 + 100{,}000(P/A,6\%,n)$
$(P/A,6\%,n) = 600{,}000/100{,}000 = 6.0$
By interpolation, $n_p = 7.67$ years

Alternative B: $0 = -600{,}000 + P_g$
$= -600{,}000 + 15{,}000 \{1-(1.2/1.06)^n]/ (0.06 - 0.2)\}$
$= -600{,}000 - 107{,}142 [1-(1.13207)^n]$

n	Result
15	$-18,396
16	$ 72,566

By interpolation, $n_p = 15.2$ years

Select A for a shorter payback at 6%.

Present worth analysis:
Determine PW over LCM of 16 years.

PW_A = -600,000 [1 + (P/F,6%,8] + 100,000(P/A,6%,16)
  = -600,000[1 + 0.6274] + 100,000(10.1059)
  = $34,150

PW_B = -600,000 + 15,000 {[1-(1.2/1.06)^{16}]/ (0.06 - 0.2)}
  = -600,000 + 15,000 {44.842144}
  = $72,632

Select B.

Difference is due to fact that PW analysis considers the increasing revenue estimates beyond the payback period. Shorter-lived alternatives are usually favored by the payback analysis.

5.30 Set up a PW relation, then use trial-and-error or spreadsheet for n_p.

$$0 = -10,000 + 1700(P/A,8\%,n) + 900(P/F,8\%,n)$$

for n = 7:   0 = -10,000 + 1700(5.2064) + 900(0.5835) = $-624
for n = 8:   0 = -10,000 + 1700(5.7466) + 900(0.5403) = $+255
  n_p = 7.7 years

Spreadsheet: NPER(8%,1700,-10000,900) displays 7.7 years

Don't purchase.

5.31 Payback analysis is best used as a supplemental analysis tool because it only determines the time necessary to recover the initial investment and a stated return (i ∃ 0%). Therefore, payback does not recognize cash flows beyond the payback period. An alternative with increasing cash flows that makes the rate of return increase may not be selected when another alternative has a shorter payback period.

5.32 Acquisition phase: Includes first 4 cost categories. In $ million:

PW_A = 2.4(P/F,6%,1) + 1.7(P/F,6%,2) + 5.9(P/F,6%,3)
  = 2.4(0.9434) + 1.7(0.8900) + 5.9(0.8396)
  = $8.7308

Operations phase: Includes last 7 categories plus all personnel costs.

$$\begin{aligned}PW_O &= 0.2(P/F,6\%,2) + 1.6(P/F,6\%,3) + 11.1(P/F,6\%,4) + 6.1(P/F,6\%,5) \\ &\quad + 1.37(P/A,6\%,19)(P/F,6\%,5) + 2.5(P/F,6\%,10) + 3.5(P/F,6\%,18) \\ &\quad + \text{personnel costs} \\ &= 0.2(0.8900) + 1.6(0.8396) + 11.1(0.7921) + 6.1(0.7473) \\ &\quad + 1.37(11.1581)(0.7473) + 2.5(0.5584) + 3.5(0.3503) \\ &\quad + \text{personnel costs} \\ &= \$28.9179 + \text{personnel costs}\end{aligned}$$

Personnel time and costs: In $ million

Year	4	5-7	8-10	11-20	21-25
$ for domestic	4.50	3.00	2.25	3.00	3.75
$ for int'l	12.00	8.00	4.00	4.40	4.80
Total	16.50	11.00	6.25	7.40	8.55

$$\begin{aligned}PW_{PC} &= 16.5(P/F,6\%,4) + 11.0(P/A,6\%,3)(P/F,6\%,4) \\ &\quad + 6.25(P/A,6\%,3)(P/F,6\%,7) + 7.4(P/A,6\%,10)(P/F,6\%,10) \\ &\quad + 8.55(P/A,6\%,5)(P/F,6\%,20) \\ &= 16.5(0.7921) + 11.0(2.6730)(0.7921) + 6.25(2.6730)(0.6651) \\ &\quad + 7.4(7.3601)(0.5584) + 8.55(4.2124)(0.3118) \\ &= 13.0697 + 23.2901 + 11.1113 + 30.4131 + 11.2298 \\ &= \$89.1140\end{aligned}$$

Total operations phase cost is:

$$PW_O = 28.9179 + 89.1140 = \$118.0319$$

Total LCC is:

$$\begin{aligned}PW_T &= PW_A + PW_O \\ &= 8.7308 + 118.0319 \\ &= \$126.7627 \qquad (\$126{,}762{,}700)\end{aligned}$$

5.33 $$\begin{aligned}PW_A &= -750{,}000 - 300{,}000(P/F,0.75,120) - (6000 + 2000)(P/A,0.75\%,240) \\ &= -750{,}000 - 300{,}000(0.4079) - 8000(111.1450) \\ &= \$-1{,}761{,}530\end{aligned}$$

$$\begin{aligned}PW_B &= -1{,}100{,}000 - (3000 + 1000)(P/A,0.75\%,240) \\ &= -1{,}100{,}000 - 4{,}000(111.1450) \\ &= \$-1{,}544{,}580\end{aligned}$$

Select Proposal B.

5.34 PW_A = -145,000 – 185,000(P/F,10%,1) – 80,000(P/A,10%,10)
       = -145,000 – 185,000(0.9091) – 80,000(6.1446)
       = $-804,752

   PW_B = -55,000 – 30,000(P/F,10%,1) – 120,000(P/A,10%,10)
       = -55,000 – 30,000(0.9091) – 120,000(6.1446)
       = $-819,625

   PW_C = -150,000(P/A,10%,10)
       = -150,000(6.1446)
       = $-921,690

   Select Alternative A.

5.35 $1200 = $\frac{(V)(0.06)}{2}$
       V = $40,000

5.36 $1500 = $\frac{(50,000)(b)}{4}$
       b = 12% per year, payable quarterly

5.37 I = $\frac{50,000(0.10)}{2}$
       = $2500 every 6 months

   PW = 2500(P/A,6%,40) + 50,000(P/F,6%,40)
       = 2500(15.0463) + 50,000(0.0972)
       = $42,476

5.38 I = $\frac{(10,000,000)(0.08)}{4}$
       = $200,000 per quarter

   PW = 200,000(P/A,3%,80) + 10,000,000(P/F,3%,80) – 250,000
       = 200,000(30.2008) + 10,000,000(0.0940) – 250,000
       = $6,730,160

5.39 I = $\frac{(10,000)(0.10)}{4}$
       = $250

   PW = 250(P/A,4%,60) + 10,000(P/F,4%,60)
       = 250(22.6235) + 10,000(0.0951)
       = $6607

5.40 (a) Cost of bond now is $50,000, because bond interest rate and market interest rate are the same. In 15 years, there are 5 years of interest left, so sales price will be the PW then (PW_{15}).

$$I = \frac{(50,000)(0.14)}{2}$$
$$= \$3500 \text{ every 6 months}$$

PW_{15} = 3500(P/A,4%,10) + 50,000(P/F,4%,10)
= 3500(8.1109) + 50,000(0.6756)
= $62,168

(b) F is the amount earned from bond interest payments + amount received from sale of bond (in part a).

F = 3500(F/A,5%,30) + 62,168
= 3500(66.4388) + 62,168
= $294,704

FE Review Problems

5.41 PW_y = -66,000 − 15,000(P/A,10%,4) + 23,000(P/F,10%,4)
= -66,000 − 15,000(3.1699) + 23,000(0.6830)
= $-97,840

Answer is (c)

5.42 PW_x = -42,000 − 42,000(P/F,10%,2) − 20,000(P/A,10%,4)
    + 10,000(P/F,10%,2) + 10,000(P/F,10%,4)
= -42,000 − 42,000(0.8264) − 20,000(3.1699) + 10,000(0.8264)
    + 10,000(0.6830)
= $-125,013

Answer is (c)

5.43 First find AW then divide by i.

AW_Y = -66,000(A/P,10%,4) − 15,000 + 23,000(A/F,10%,4)
= -66,000(0.31547) − 15,000 + 23,000(0.21547)
= $-30,865

CC = P = -30,865/0.10
= $-308,650

Answer is (c)

5.44 Convert the $5,000 amounts into an A value and then divide by i.

CC = P = -1,000/0.10 – 5,000(A/F,10%,5)/0.10
= -10,000 – 5,000(0.16380)/0.10
= $-18,190

Answer is (a)

5.45 Answer is (c)

5.46 Answer is (b)

5.47 CC = P = -200,000 – 30,000/0.10
= $-500,000

Answer is (c)

5.48 First find AW then divide by i.

A = -20,000(A/P,10%,8) – 6000 + 5000(A/F,10%,8)
= -20,000(0.18744) – 6000 + 5000(0.08744)
= $-9312

CC = PW = -9,312/0.10
= $-93,120

Answer is (c)

5.49 P = -30,000 – 1000/0.01 – 5000(A/F,1%,60)/0.01
= -30,000 – 100,000 – 5000(0.01224)/0.01
= $-136,120

Answer is (b)

Extended Exercise Solution

Questions 1, 3 and 4:

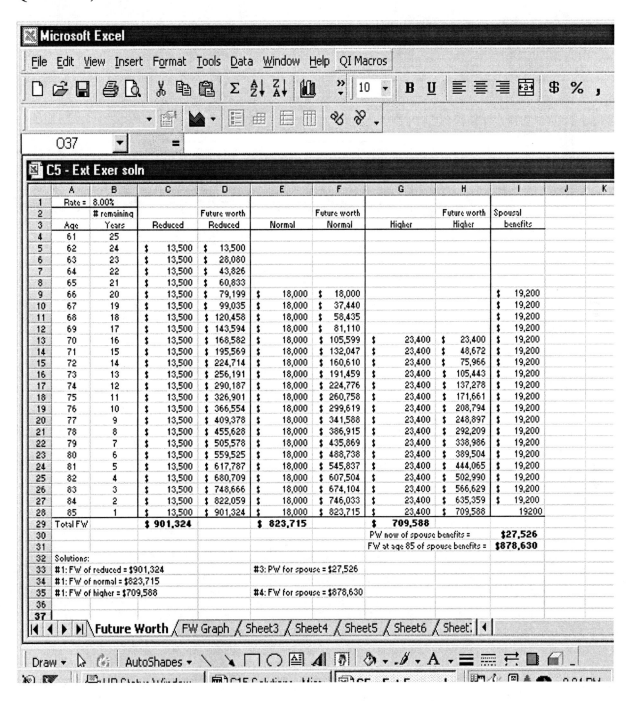

Question 2:

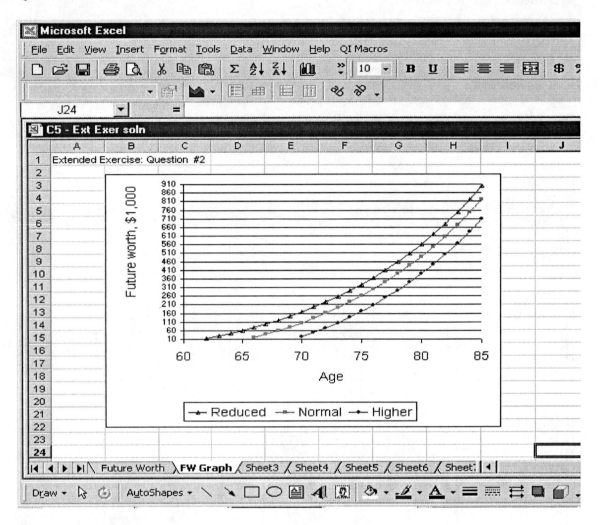

Case Study Solution

1. Set first cost of toilet equal to monthly savings and solve for n:

    $[(115.83 - 76.12) + 50](A/P, 0.75\%, n) = 2.1(0.76 + 0.62)$
    $89.71(A/P, 0.75\%, n) = 2.898$
    $(A/P, 0.75\%, n) = 0.03230$

    From 0.75% interest table, n is between 30 and 36 months

    By interpolation, n = 35 months or 2.9 years

2. If the toilet life were to decrease by 50% to 2.5 years, then the homeowner would not breakeven at any interest rate (2.6 years is required at 0% and longer times would be required for i > 0%). If the interest rate were to increase by more than 50% (say from 9% to 15%), the payback period would increase from 2.9 years (per above solution) to a little less than 3.3 years (from 1.25% interest table). Therefore, the payback period is much more sensitive to the toilet life than to the interest rate.

3. cost/month = 76.12 (A/P, 0.5%, 60)
    = 76.12 (0.01933)
    = $1.47

    CCF/month = 2.1 + 2.1
    = 4.2

    cost/CCF = 1.47/4.2
    = $0.35/CCF or $0.47/1000 gallons (vs $0.40/1000 gallons at 0% interest)

4. (a) If 100% of the $115.83 cost of the toilet is rebated, the cost to the city at 0% interest is

    $$c = \frac{115.83}{(2.1 + 2.1)(12)(5)}$$

    = $0.46/CCF or $0.61/1000 gal   (vs $0.40/1000 gal at 75% rebate)

    This is still far below the city's cost of $1.10/1000 gallons. Therefore, the success of the program is not sensitive to the percentage of cost rebated.

    (b) Use the same relation for cost/month as in question 3 above, except with varying interest rates, the values shown in the table below are obtained for n = 5 years:

Interest Rate, %	4	6	8	10	12	15
$ / CCF	0.33	0.35	0.37	0.39	0.40	0.43
$ / 1000 gallons	0.45	0.47	0.49	0.51	0.54	0.58

The results indicate that even at an interest rate of 15% per year, the cost at $0.58/1000 gallons is significantly below the city's cost of $1.10/1000 gallons. Therefore, the program's success is not sensitive to interest rates.

(c) Use the same equation as in question 3 above with i = 0.5% per month and varying life values.

Life, years	2	3	4	5	6	8	10	15	20
$ / CCF	0.80	0.55	0.43	0.35	0.30	0.24	0.20	0.15	0.13
$ / 1000 gallons	1.07	0.74	0.57	0.47	0.40	0.32	0.27	0.20	0.17

For a 2-year life and an interest rate of a nominal 6% per year, compounded monthly, the cost of the program is $1.07/1000 gallons, which is very close to the savings of $1.10/1000 gallons. But the cost decreases rapidly as life increases.

If further sensitivity analysis is performed, the following results are obtained. At an interest rate of 8% per year, the costs and savings are equal. Above 8% per year, the program would not be cost effective for a 2-year toilet life at the 75% rebate level. When the rebate is increased to 100%, the cost of the program exceeds the savings at all interest rates above 4.5% per year for a toilet life of 3 years.

These calculations reveal that at very short toilet lives (2-3 years), there are some conditions under which the program will not be financially successful. Therefore, it can be concluded that the program's success is mildly sensitive to toilet life.

Chapter 6 – Annual Worth Analysis
Solutions to end of chapter exercises

Problems

6.1 The AW method assumes: (1) The services provided are needed forever; (2) The alternatives will be repeated exactly the same in succeeding life cycles, and (3) All cash flows will have the same values in succeeding life cycles (or more correctly, they will change only by the same amount as the inflation or deflation rate).

6.2 $AW_A = -10,000 (A/P, 15\%, 2) + 7000$
$= -10,000 (0.61512) + 7000$
$= \$848.80$

$AW_B = -10,000(A/P,15\%,4) + 7000(P/F,15\%,1)(A/P,15\%,4) - 3000$
$(P/F,15\%,2)(A/P,15\%,4) + 7000(P/A,15\%,2)(P/F,15\%,2)(A/P,15\%,4)$
$= -10,000 (0.35027) + 7000(0.8696)(0.35027) - 3000(0.7561)(0.35027) +$
$7000(1.6257)(0.7561)(0.35027)$
$= \$848.79 \quad (1¢ \text{ round-off error})$

The AW's are the same because the cash flow for Alt B is the cash flow for A that has been repeated for one more life cycle.

6.3 (a) Using Eq. [6.3]: $CR = -[50,000(A/P,12\%,4) - 10,000(A/F,12\%,4)]$
$= -[50,000(0.32923) - 10,000(0.20923)]$
$= \$-14,369.20$

Using Eq. [6.4]: $CR = -[(50,000 - 10,000)(A/P,12\%,4) + 10,000 (0.12)]$
$= -[40,000 (0.32923) + 1200]$
$= \$-14,369.20$

(b) Capital recovery is the equivalent annual amount that must be recovered by the alternative to return the initial cost and a 12% rate of return over the 4-year life.

6.4 $AW_A = -80,000(A/P,18\%,3) - 30,000 + 15,000(A/F,18\%,3)$
$= -80,000(0.45992) - 30,000 + 15,000(0.27992)$
$= \$-62,595$

$AW_B = -120,000(A/P,18\%,2) - 8000 + 40,000(A/F,20\%,2)$
$= -120,000(0.63872) - 8000 + 40,000(0.45872)$
$= \$-66,298$

Use Method A.

6.5　$AW_{20\%}$ = -100(A/P,10%,10) – [80–0.20(80)]
　　　　　= -100(0.16275) – 64
　　　　　= $-80.28

　　　$AW_{35\%}$ = -240(A/P,10%,15) – [80 – 0.35(80)]
　　　　　= -240(0.13147) – 52
　　　　　= $-83.55

　　Washers with 20% increase are more economical.

6.6　Cost of tap water per 8 ounce glass = $\dfrac{1.90 * 8 * 100}{1000 * 128}$

　　　　　　　　　　　　　　　= 0.012¢ per glass

　　Move daily water costs to end-of-month costs:

　　Bottled water cost/month = 0.28 * 5 * 30
　　　　　　　　　　　　　= $42 per month

　　Tap water cost/month = 0.00012 * 5 * 30
　　　　　　　　　　　= $0.018 per month

　　　　$AW_{Bottled}$ = 42(F/A,0.5%,12)
　　　　　　　　= 42(12.3356)
　　　　　　　　= $518.10 per year

　　　　AW_{Tap} = 0.018 (F/A,0.5%,12)
　　　　　　　= 0.018(12.3356)
　　　　　　　= $0.22 per year
　　　　　　　= 22¢ per year

6.7　$AW_{new\ line}$ = -2.5 million(A/P,20%,5)
　　　　　　= -2.5 million(0.33438)
　　　　　　= $-835,950

　　AW_{lean} = -1.4 million(A/P,20%,3)
　　　　　= -1.4 million(0.47473)
　　　　　= $-664,622

　　Adopt lean manufacturing techniques.

6.8　AW_{single} = [-4000 – 4000(P/A,12%,2)](A/P,12%,5)
　　　　　　= [-4000 – 4000(1.6901)](0.27741)
　　　　　　= $-2,985

$$AW_{site} = -10,000(A/P,12\%,5)$$
$$= -10,000(0.27741)$$
$$= \$-2774$$

Buy the site license.

6.9 $AW_{variable} = -45,000(A/P,15\%,8) - 31,000 - 12,000(P/F,15\%,5)(A/P,15\%,8)$
           $+ 10,000(A/F,15\%,8)$
        $= -45,000(0.22285) - 31,000 - 12,000(0.4972)(0.22285)$
           $+ 10,000(0.07285)$
        $= \$-41,629$

$AW_{dual} = -24,000(A/P,15\%,6) - 35,000 - 6000[(P/F,15\%,2) + (P/F,15\%,4)]$
       $(A/P,15\%,6) + 8000(A/F,15\%,6)$
    $= -24,000(0.26424) - 35,000 - 6,000[0.7561 + 0.5718](0.26424) + 8000$
       (0.11424)
    $= \$-42,533$

Purchase the variable speed machine.

6.10 (a) $AW_X = -150,000(A/P,16\%,3) - [80,000 + 3000(A/G,16\%,3)] + 20,000$
          $(A/F,16\%,3)$
       $= -150,000(0.44526) - [80,000 + 3000(0.9014)] + 20,000(0.28526)$
       $= \$-143,788$

$AW_Y = -240,000(A/P,16\%,5) - [60,000 + 2500(A/G,16\%,5)] + 32,000$
       $(A/F,16\%,5)$
    $= -240,000(0.30541) - [60,000 + 2500(1.7060)] + 32,000(0.14541)$
    $= \$-132,910$

$AW_Z = -310,000(A/P,16\%,6) - [48,000 + 1000(A/G,16\%,6)] + 36,000$
       $(A/F,16\%,6)$
    $= -310,000(0.27139) - [48,000 + 1000(2.0729)] + 36,000(0.11139)$
    $= \$-130,194$

Select alternative Z.

(b) Use the LCM of 30 years.

$PW_X = AW_X(P/A,16\%,30)$
     $= -143,788(6.1772)$
     $= \$-888,207$

$PW_Y = -132,910(P/A,16\%,30)$
     $= -132,910(6.1772)$
     $= \$-821.012$

Chapter 6

$$PW_Z = -130{,}194(P/A,16\%,30)$$
$$= -130{,}194(6.1772)$$
$$= \$-804{,}234$$

Select alternative Z.

6.11 $AW_{purchase} = -150{,}000(A/P,20\%,8) + 65{,}000(A/F,20\%,8) + 10{,}000$
   $= -150{,}000(0.26061) + 65{,}000(0.06061) + 10{,}000$
   $= \$-25{,}152$

$AW_{lease} = -[30{,}000 + 30{,}000(P/A,20\%,7)](A/P,20\%,8)$
   $= -[30{,}000 + 30{,}000(3.6046)](0.26061)$
   $= \$-36{,}000$

Purchase the clamshell.

6.12 $AW_{patch} = -[(300 * 700) + 24{,}000](A/P,9\%,2) - 5{,}000$
   $= [234{,}000](0.56847) - 5000$
   $= \$-138{,}022$

$AW_{resurface} = -850{,}000(A/P,9\%,10) - [2000(P/A,9\%,7)](P/F,9\%,3)(A/P,9\%,10)$
   $= -850{,}000(0.15582) - [2000(5.0330)](0.7722)(0.15582)$
   $= \$-133{,}658$

Resurface the road.

6.13 (a) $AW_{petroleum} = -250{,}000(A/P,18\%,6) - 130{,}000 + 400{,}000 + 50{,}000(A/F,18\%,6)$
   $= -250{,}000(0.28591) + 270{,}000 + 50{,}000 (0.10591)$
   $= \$203{,}818$

$AW_{inorganic} = -110{,}000(A/P,18\%,4) - 65{,}000 + 270{,}000 + 20{,}000(A/F,18\%,4)$
   $= -110{,}000(0.37174) + 205{,}000 + 20{,}000(0.19174)$
   $= \$167{,}943$

Select petroleum-based alternative.

(b) Both are acceptable since AW > 0 at the MARR.

6.14 $AW_{remodel} = -10{,}000{,}000(A/P,10\%,10) - 2{,}000{,}000(P/F,10\%,4)(A/P,10\%,10)$
   $+ [70{,}000(P/F,10\%,1) + 65{,}000(P/F,10\%,2) + 60{,}000(P/A10\%,8)(P/F,10\%,2)](A/P,10\%,10)$
   $= -10{,}000{,}000(0.16275) - 2{,}000{,}000(0.6830)(0.16275) + [70{,}000(0.9091) + 65{,}000(0.8264) + 60{,}000(5.3349)(0.8264)] 0.16275)$
   $= \$-1{,}787{,}666$

$$AW_{new} = -40,000,000(A/P,10\%,40) + 190,000$$
$$= -40,000,000(0.10226) + 190,000$$
$$= \$-3,900,400$$

Remodel the existing arena.

6.15 $AW = -400,000(0.10) - 50,000$
    $= -40,000 - 50,000$
    $= \$-90,000$ per year

6.16 $AW = -100,000(0.20) - 30,000 - 50,000(A/F,20\%,5)$
    $= -20,000 - 30,000 - 50,000(0.13438)$
    $= \$-56,719$ per year

6.17 First find P in year 10 for the $10 million annual amounts and then use the A/F factor to find A ($ in million):

$$P_{10} = \frac{-10}{0.15}$$
$$= \$-66.667$$

$A = -66.667(A/F,15\%,11)$
  $= -66.667(0.04107)$
  $= \$-2,738,000$ per deposit

6.18 The perpetual uniform annual worth is the AW on one life cycle:

$AW = -40,000(A/P,20\%,3) - 24,000 + 6000(A/F,20\%,3)$
   $= -40,000(0.47473) - 24,000 + 6000(0.27473)$
   $= \$-41,341$

6.19 First find PW and then convert to AW:

$PW = -[6000(P/A,10\%,13) + 1000(P/G,10\%,13)](P/F,10\%, 2) - 18,000$
    $(P/A,10\%,25) (P/F,10\%,15)$
   $= -[6000(7.1034) + 1000(33.3772)](0.8264) - 18,000(9.0770)(0.2394)$
   $= \$-101,919$

$AW = -101,919(A/P,10\%,40)$
   $= -101,919(0.10226)$
   $= \$-10,422$

Spreadsheet solution: Enter costs in cells $A_1 - A_{40}$: 0.0 – 6000, 7000, etc. to -18,000 in A_{15}, and -18,000 in $A_{16} - A_{40}$. (There is no cash flow in year 0.)

NPV(10%,A1,A40) = $-101,919
PMT(10%,40,-NPV_cell) = $-10,422

6.20 Find F in year 9 and multiply by i:

P_{-1} = -[60(P/A,15%,5) – 10(P/G,15%,5)]
   = -[60(3.3522) – 10(5.7751)]
   = $-143,381

F_9 = -143,381(F/P,15%,10)
   = -143,381(4.0456)
   = $-580,062

AW = -580,062(0.15)
   = $-87,009

6.21 (a) $AW_{inhouse}$ = -40(A/P,20%,10) – 5 + 14 + 7(A/F,20%,10)
      = -40(0.23852) + 9 + 7(0.03852)
      = -0.27116   ($-271,160)

   $AW_{license}$ = -2(0.20) – 0.2 + 6
      = -0.4 + 5.8
      = 5.4   ($5,400,000)

License the production.

(b) Only the license option is acceptable at i = 20% per year.

6.22 AW_X = -60,000(A/P,1%,60) – 30,000 +7000(A/F,1%,60)
      = -60,000(0.02224) – 30,000 + 7000(0.01224)
      = $-31,249 per month

   AW_Y = -300,000(A/P,1%,240) – 20,000 +25,000(A/F,1%,240)
      = -300,000(0.01101) – 20,000 + 25,000(0.00101)
      = $-23,278 per month

   AW_Z = -900,000(0.01) – 15,000 – 60,000(A/F,1%,48)
      = -9000 – 15,000 – 60,000(0.01633)
      = $-24,980 per month

Select Alt Y.

FE Review Problems

6.23 Answer is (d)

6.24 $AW_A = -50{,}000(A/P,10\%,3) - 20{,}000 + 10{,}000(A/F,10\%,3)$
      $= -50{,}000(0.40211) - 20{,}000 + 10{,}000(0.30211)$
      $= \$-37{,}084$

Answer is (b)

6.25 $AW_B = -80{,}000(A/P,10\%,6) - 10{,}000 + 25{,}000(A/F,10\%,6)$
      $= -80{,}000(0.22961) - 10{,}000 + 25{,}000(0.12961)$
      $= \$-25{,}129$

Answer is (a)

6.26 Perpetual AW is same as AW for one life cycle:

$AW_A = -50{,}000(A/P,10\%,3) - 20{,}000 + 10{,}000(A/F,10\%,3)$
$= -50{,}000(0.40211) - 20{,}000 + 10{,}000(0.30211)$
$= \$-37{,}084$

Answer is (d)

6.27 $AW_F = -200{,}000(A/P,10\%,10) - 50{,}000 + 120{,}000 + 25{,}000(A/F,10\%,10)$
      $= -200{,}000(0.16275) + 70{,}000 + 25{,}000(0.06275)$
      $= \$-39{,}019$

Answer is (c)

6.28 $AW_G = -1{,}000{,}000\,(0.10) - 10{,}000 + 140{,}000$
      $= -100{,}000 + 130{,}000$
      $= \$30{,}000$

Answer is (b)

6.29 $AW = -70{,}000(0.10) - 100{,}000(P/F,10\%,7)(0.10) - \dfrac{10{,}000(P/F,10\%,9)(0.10)}{0.10}$
      $= -7000 - 100{,}000(0.5132)(0.10) - 10{,}000(0.4241)$
      $= \$-16{,}373$

Answer is (a)

6.30 $AW = -5{,}000{,}000(0.10) - 60{,}000 - 100{,}000(A/F,10\%,5)$
      $= -500{,}000 - 60{,}000 - 100{,}000(0.16380)$
      $= \$-576{,}380$

Answer is (a)

Case Study Solution

1. Spreadsheet and chart are below. Revised costs and savings are in columns F-H.

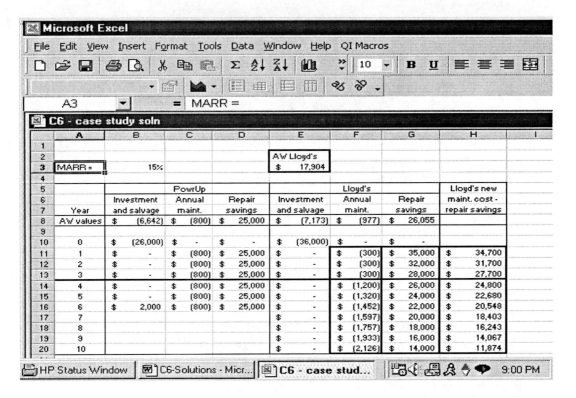

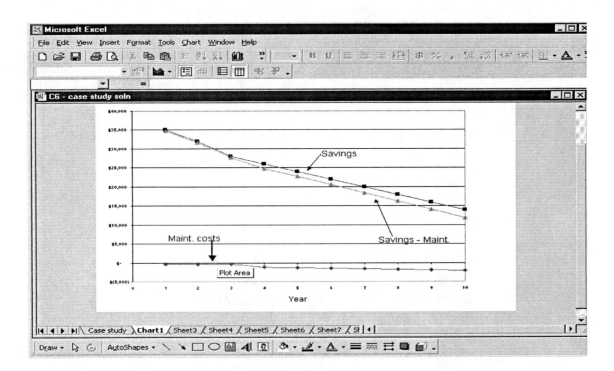

2. In cell E3, the new AW = $17,904. This is only slightly larger than the PowrUp AW = $17,558. Marginally select Lloyd's.

3. New CR = $-7172, which is an increase from the $-7025 previously displayed in cell E8.

Chapter 7 – Rate of Return Analysis: Single Alternative
Solutions for the end of chapter exercises

Problems

7.1 At –100%, the entire balance is lost, so there is no unrecovered balance below this point and, therefore, there can be no rate of return.

7.2 When interest is charged on the unrecovered balance, A is:
$A = 5000(A/P,10\%,5)$
  $= 5000(0.26380)$
  $= \$1319$

Amount owed immediately before first payment is:
$F_{1before} = 5000(F/P,10\%,1)$
     $= 5000(1.1000)$
     $= \$5500$

Amount owed after first payment is:
$F_{1after} = 5500 - 1319$
     $= \$4181.00$

Interest in first payment = $\$5000(0.10) = \$500$.

Amount owed immediately before second payment is:
$F_{2before} = 4181(F/P,10\%,1)$
     $= 4181(1.1000)$
     $= \$4599.10$

Interest in second payment = $4599.10 - 4181.00$
                = $\$418.10$

When interest is charged on loan principal, interest each year is $5000(0.10) = \$500$.

Difference = $500 - 418.10$
       = $\$81.90$

7.3 Loan payment by A-1 mortgage scenario is:
Loan payment = Principal reduction + interest
       = $\dfrac{10{,}000}{4} + 10{,}000(0.10)$
       = $2500 + 1000$
       = $\$3500$

Use A/P or P/A factor to find i. Using P/A,
$$10,000 = 3500(P/A,i,4)$$
$$(P/A,i,4) = 2.8571$$
$$i = 14.96\% \quad \text{(RATE or IRR function)}$$

7.4 (a) Use P/F or F/P factor and solve for i. Using F/P:
$$58 = 5.5(F/P,i,24)$$
$$(F/P,i,24) = 10.5454$$
$$i = 10.31\% \quad \text{(RATE or IRR function)}$$

(b) Solve for F in year 2008 using $i = 10.31\%$
$$F = 58 \, (F/P, 10.31\%, 9)$$
$$= 58 \, (2.4184)$$
$$= 140 \text{ liters per person}$$

7.5 $0 = -25,000 + (-18,000 + 27,000)(P/A,i,5)$
$(P/A,i,5) = 2.7777$
$i^* = 23.44\%$ (RATE or IRR function)

7.6 $0 = -126,000 + (-49,000 + 88,000)(P/A,i,8) + 33,000(P/F,i,8)$
$i^* = 27.57\%$ (RATE or IRR function)

7.7 $0 = -10 + (2-3)(P/F,i,1) + (4-4)(P/F,i,2) + (5-3)(P/F,i,3) + (6-3)(P/F,i,4)$
$\quad + (9-3)(P/F,i,5) + (10-4)(P/F,i,6) + (11-4)(P/F,i,7)$
$i^* = 15.74\%$ (RATE or IRR function)

7.8 NCF in years 0 – 5 in $1000 units: -311, 126, 126, 126, 126, 218

$0 = -220,000 - 15,000 - 76,000 + (2-1.05)(120,000)(P/A,i,5) + 12,000(P/A,i,5)$
$\quad + 92,000(P/F,i,5)$

$i^* = 33.19\%$ per year   (RATE or IRR function)

7.9 $0 = -30 - 28(P/F,i,1) - 24(P/F,i,2) - 15(P/F,i,3) + 3(P/F,i,4)$
$\quad + 17(P/F,i,5) + 30(P/A,i,5)(P/F,i,5)$

Solve by trial and error or using the IRR function:
$i^* = 9.31\%$ per year   (IRR function)

7.10 NCF for month 0: $-90,000; months 1 – 24: $7440

$0 = -90,000 - 0.012(5000)(P/A,i,24) + 0.01(5000)(150)(P/A,i,24)$

Solve by trial and error or spreadsheet:
$i^* = 6.40\%$ per month   (RATE or IRR function)

7.11 NCF in $1000 units for years 0 – 5: -210, -150, 220, 280, 340, 400

$0 = -210 - 150(P/F,i,1) + [220(P/A,i,4) + 60(P/G,i,4)](P/F,i,1)$

Solve by trial and error or using IRR function:
$i^* = 48.43\%$ per year (IRR function)

7.12 $0 = -450,000 - 50,000(P/F,i,1) - 40,000(P/F,i,2) - 30,000(P/F,i,3) + 10,000(P/F,i,4) + 30,000(P/F,i,5) + 50,000(P/A,i,19)(P/F,i,5)$

Solve by trial and error or spreadsheet:
$i^* = 4.16\%$ per month (IRR function)
$= 4.16 * 12$
$= 49.92\%$ per year (nominal)

7.13 NCF in $1000 units: year 0: $-750; years 1–10: $0; year 11: $300;...; year 15: $500

$0 = -750,000 + [300,000(P/A,i,5) + 50,000(P/G,i,5)](P/F,i,10)$

Solve by trial and error or spreadsheet:
$i^* = 7.73\%$ per year (IRR function)

7.14 Rate of return equations are more time consuming because they usually require a trial and error solution.

7.15 $0 = -10,000,000 + \dfrac{(150)(10,000)}{i}(P/F,i,9)$

Solve for i^* by trial and error:

Try i = 8%: $0 = -10,000,000 + \dfrac{1,500,000}{0.08}(P/F,8\%,9)$
$0 = -10,000,000 + 8,750,000(0.5002)$
$0 > \$-621,250$

Try i = 7.5%: $0 = -10,000,000 + 20,000,000(0.52158)$
$0 < \$431,600$

$i^* = 7.705\%$ (interpolation)

7.16 $0 = -5,000,000 + 200,000(P/A,i,5) + \dfrac{1,000,000}{i}(P/F,i,5)$

Solve for i* by trial and error:
Try i = 12%: $0 = -5,000,000 + 200,000(3.6048) + \dfrac{1,000,000}{0.12}(0.5674)$
$= -5,000,000 + 720,960 + 4,728,333$
$0 < \$449,293$

Try i = 14%:   0 > $-603,380

i* = 12.77%   (interpolation)

7.17 The PW method assumes that all net positive cash flows for a project are reinvested at the MARR. The ROR method makes the same reinvestment assumption but at the i* rate. Unreasonable decisions can result when MARR and i* are not close to each other.

7.18 A nonconventional cash flow series has more than one sign change in the net cash flow series.

7.19 The maximum number of roots that will balance a rate of return equation is equal to the number of sign changes in the net cash flow series.

7.20 According to Norstrom's criterion, there is only one positive value for a rate-of-return equation when the cumulative cash flow series (1) starts out negative, and (2) has only 1 sign change.

7.21 (a) There are two sign changes in net cash flow. Therefore, there are two possible rate of return values.

(b) The cumulative cash flow is negative for all years. There are no positive i values. (i* = -13.37% by the IRR function.)

7.22 (a) Since there is only one sign change in net cash flow, there is only one rate of return value.

(b) $0 = -17{,}000 - 20{,}000(P/F,i,1) + 4000(P/F,i,2) + 50{,}000(P/F,i,3)$

Solve by trial and error spreadsheet:
i* = 17.07% per year   (IRR function)

7.23 There are 3 sign changes in net cash flows but only one change in cumulative cash flow that starts negatively. Therefore, there is a unique rate of return value:

$0 = -33{,}000 - 3000(P/F,i,1) + 67{,}000(P/F,i,2) - 15{,}000(P/F,i,3) + 9000(P/F,i,4)$

Solve by trial and error or spreadsheet:
i* = 31.36% per year   (IRR function)

7.24

Year	0	1	2	3	4	5	6
Cash flow, $	5000	-10,100	500	2000	2000	2000	2000
Cumulative CF, $	5000	-5100	-4600	-2600	-600	1400	3400

Two sign changes in cash flow and cumulative cash series. Up to 2 roots are present.

$i_1^* = 34.06\%$
$i_2^* = 61.37\%$

Excel worksheet with (1) plot of PW versus i, and (2) i* using IRR with guess operator is shown.

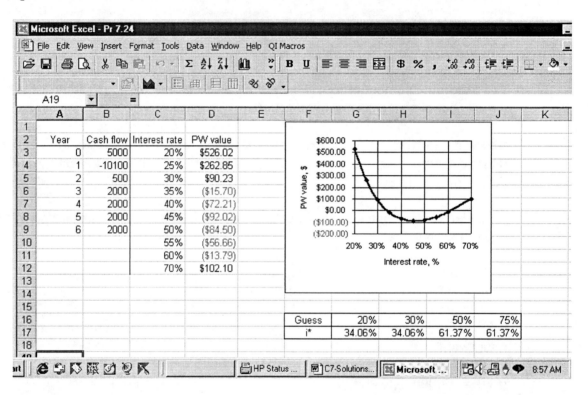

7.25

Year	0	1	2	3	4	5
Cum. CF,$	-5000	0	0	0	15,000	0

Two sign changes in cash flow series. Cumulative cash flows have 2 changes (in theory).

$i_1^* = 0.0\%$
$i_2^* = 31.61\%$

Excel worksheet with (1) PW versus i graph, and (2) i* using IRR with guess operator is shown.

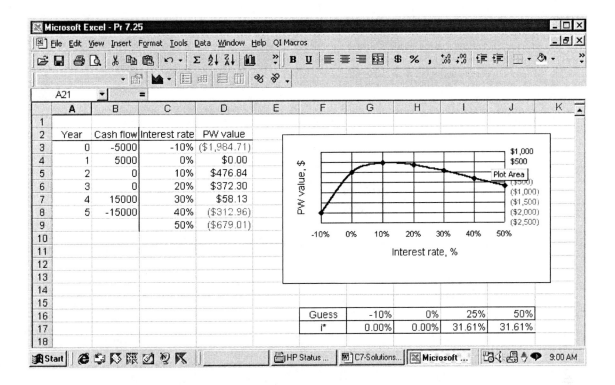

7.26 (a) There is only one sign change, so the possible number of rate of return values is one.

(b) The table below shows that the number of shares purchased each year was 100.

Year	Bonus	Share Price	No. Shares Purchased
1	4,000	40	100
2	5,000	50	100
3	6,000	60	100
4	7,000	70	100
5	8,000	80	100
6	9,000	90	100
7	10,000	100	100
8	11,000	110	100
9	12,000	120	100
10	13,000	130	100

The rate of return equation is:

$$0 = -[4000((P/A,i,10) + 1000(P/G,i,10)] + 1000(130)(P/F,i,10)$$

Solve by trial and error or spreadsheet:
$i^* = 11.40\%$ (IRR function)

(c) There is no indication of multiple roots, so c = iN = i*. If Eq. [7.6] is applied, all F values are negative except the last one, so iN is used in all equations. Therefore, the composite rate of return is the same as the internal rate, iN = i* = 11.40%.

7.27 (a) Norstrom's criterion has one sign change so a single i* is present.

$$0 = -65 + 20(P/F,i,1) + 44(P/F,i,2) + 68(P/F,i,3) - 12(P/F,i,4)$$

Solve by trial and error or spreadsheet:
i* = 32.98%   (IRR function)

(b) Apply net-investment procedure, since reinvestment is not at the i* = 32.98%.

$F_0 = -65$ $F_0 < 0$; use i′
$F_1 = -65(1+i′) + 20$ $F_1 < 0$; use i′
$F_2 = F_1(1+i′) + 44$ $F_2 < 0$; use i′
$F_3 = F_2(1+i′) + 68$ $F_3 > 0$; use c (F_3 must be > 0 because last term is negative.)
$F_4 = F_3(1 + 0.14) - 12$

Set $F_4 = 0$ and solve for iN by trial and error.

$$0 = \{[[-65(1+i′) + 20](1+i′) + 44](1+i′) + 68\}(1+0.14) - 12$$

Try i = 30%:   0 < $6.4623
Try i = 32%:   0 < $1.0303
Try i = 33%:   0 > $-1.7667

iN = 32.37%   (interpolation)

7.28 (a) $F_0 = 3000$ $F_0 > 0$, use c = 0.12

$F_1 = 3000(1+0.12) - 2000$
    $= 1360$ $F_1 > 0$, use c = 0.12

$F_2 = 1360(1+0.12) + 1000$
    $= 2523.20$ $F_2 > 0$, use c = 0.12

$F_3 = 2523.20(1+0.12) - 6000$
    $= -3174.02$ $F_3 < 0$, use i′

$F_4 = -3174.02(1+i′) + 5000$

Set $F_4 = 0$ and solve for i′.
$0 = -3174.02(1+i′) + 5000$
iN = 57.5% per year

Chapter 7

(b) Excel chart shows that the PW versus i curve does not touch the PW = 0 line in the range –10% to 70%.

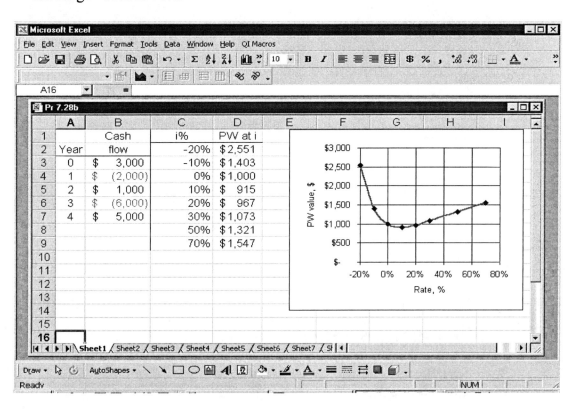

7.29 $F_0 = -5000$ $F_0 < 0$; use iN
     $F_1 = -5000(1+iN) + 5000$ $F_1 < 0$; use iN
     $F_2 = F_1(1+iN) + 0$ $F_2 < 0$; use iN
     $F_3 = F_2(1+iN) + 0$ $F_3 < 0$; use iN
     $F_4 = F_3(1+iN) + 15,000$ $F_4 > 0$ (because the last term is negative); use c
     $F_5 = F_4(1+0.15) - 15,000$

   Set $F_5 = 0$ and solve for iN.
   $0 = \{[\{[-5000(1+iN) + 5000](1+iN)\}(1+iN)](1+iN) + 15,000\}(1.15) - 15,000$

   Try i = 20%:   0 < $+263
   Try i = 21%:   0 < $+111
   Try i = 22%:   0 > $-47

   iN = 21.7%   (interpolation)

7.30 $I = \dfrac{10,000(0.08)}{2}$
      = $400 per six months

$0 = -9000 + 400(P/A,i,12) + 10,000(P/F,i,12)$
Solve for i* by trial and error or spreadsheet.

i* = 5.14% per six months   (RATE or IRR function)

Nominal i/year = 5.14 * 2 = 10.28% per year

7.31 $I = \dfrac{5,000,000(0.06)}{2}$
 = $150,000 every six months

$0 = 4,500,000 - 100,000 - 150,000(P/A,i,80) - 5,000,000(P/F,i,80)$

Solve for i* by trial and error or spreadsheet.
i* = 3.4% per six months   (RATE or IRR function)

7.32 $I = \dfrac{5000(0.06)}{4}$
 = $75 per quarter

$0 = -4000 + 75(P/A,i,48) + 5000(P/F,i,48)$

Solve for i* by trial and error or spreadsheet.
i* = 2.176% per quarter   (RATE or IRR function)

7.33 $I = \dfrac{5,000(0.12)}{4}$
 = $150 per quarter

$0 = -5000(F/P,i,32) + 150(F/A,i,32) + 5500$

Solve for i* by trial and error or spreadsheet.
i* = 3.18% per quarter   (RATE or IRR function)

7.34 $I = \dfrac{10,000,000(0.14)}{2}$
 = $700,000 every six months

$0 = -11,000,000 + 700,000(P/A,i,40) + 10,000,000(P/F,i,40)$

Solve for i* by trial and error or spreadsheet.
i* = 6.31% per six months   (RATE or IRR function)
 = 6.31 * 2 = 12.6% per year (nominal)

FE Review Solutions

Note: Use of computers with Excel is probably not allowed in the FE Exam. To shorten the time to find the correct answer for ROR problems, one approach is to try the middle two multiple-choice values first; not the highest or lowest. This should reduce the number of attempts to one or two.

7.35 By inspection, the payments are interest only with a final principal payment of $1 million after 2 years. Hence, i = 5000/1,000,000 = 0.5%, or

$$0 = 1,000,000 - 5000(P/A,i,24) - 1,000,000(P/F,i,24)$$

Solve for i* by trial and error or spreadsheet.
i = 0.50% per month   (RATE function)

Answer is (a)

7.36 $0 = -50,000 + 10,000(P/A,i,10) + 5000(P/F,i,5)$

Solve for i* by trial and error or spreadsheet.
i* = 16.38% per year   (IRR function)

Answer is (c)

7.37 There are two changes in sign in the net cash flow. Therefore, there are two possible solutions.

Answer is (c)

7.38 $0 = -40,000 + (7500 - 5000)(P/A,i,24) + 12,000(P/F,i,24)$

Solve for i* by trial and error or spreadsheet.
i* = 4.59% per month   (RATE function)

Answer is (b)

7.39 $0 = -50,000(F/P,i,4) + \dfrac{5000}{i}$

Solve for i* by trial and error or spreadsheet.
i = 7.49% per year   (trial and error)

Answer is (a)

7.40 Answer is (b)

7.41 I = $250 = 10,000(b)/4
      b = 1000/10,000 = 0.10

   Answer is (d)

7.42 0 = -9000 + 400(P/A,i,10) + 10,000(P/F,i,10)

   Solve for i* by trial and error or spreadsheet.
   i* = 5.31% per year   (RATE or IRR function)

   Answer is (c)

7.43 The rate of return will be the same as the bond interest rate of 10% per year, compounded quarterly. This is 2.5% per quarter.

   Answer is (b)

7.44 I = $\frac{10,000(0.12)}{2}$
     = $600 per six months

   0 = -11,000 + 600(P/A,i,10) + 11,000(P/F,i,10)

   Solve for i* by trial and error or spreadsheet.
   i* = 5.46% per six months   (RATE or IRR function)

   Answer is (b)

7.45 I = $\frac{3,000,000(0.12)}{2}$
     = $180,000 per six months

   0 = 2,500,000 – 180,000 (P/A, i, 40) – 3,000,000 (P/F, i, 40)

   Solve for i* by trial and error or spreadsheet.
   i = 7.29% per six months   (RATE or IRR function)

   nominal i/year = 2 * 7.29
                  = 14.58%

   Answer is (d)

Extended Exercise 1 Solution

Solution by hand:

1. Charles' payment = 5000(A/P,10%,3)
   = 5000(0.402115) (by formula)
   = $2010.57

Year (1)	Beginning unrecovered balance (2)	Interest (3) = 0.1(2)	Total amount owed (4)=(2)+(3)	Payment (5)	Ending unrecovered balance (6)=(4)+(5)
0			$-5000.00		$-5000.00
1	$-5000.00	$-500.00	-5500.00	$2010.57	-3489.43
2	-3489.43	-348.94	-3838.37	2010.57	-1827.80
3	-1827.80	-182.78	-2010.58	2010.57	-0.01*
		$-1031.72		$6031.71	

*round-off

Jeremy's payment = $2166.67

Year (1)	Beginning unrecovered balance (2)	Interest (3) = 0.1(5000)	Total amount owed (4)=(2)+(3)	Payment (5)	Ending unrecovered balance (6)=(4)+(5)
0			$-5000.00		$-5000.00
1	$-5000.00	$-500	-5500.00	$2166.67	-3333.33
2	-3333.33	-500	-3833.33	2166.67	-1666.67
3	-1666.67	-500	-2166.67	2166.67	0.00
		$-1500		$6500.01	

Plot year versus column (4) in the form of Figure 7–1 in the text.

2.
	Jeremy	Charles	More for Jeremy
Interest	$1500.00	$1031.72	$468.28
Total	6500.01	6031.71	468.30

Solution by computer:

1. The following spreadsheets have the same information as the two tables above. The x-y scatter charts are year (column A) versus total owed (column B). (The indicator lines and curves were drawn separately.)

2. The second spreadsheet shows that $468.28 more is paid by Jeremy.

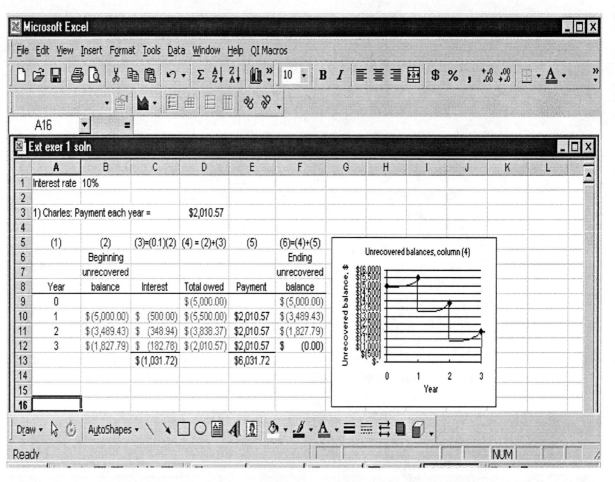

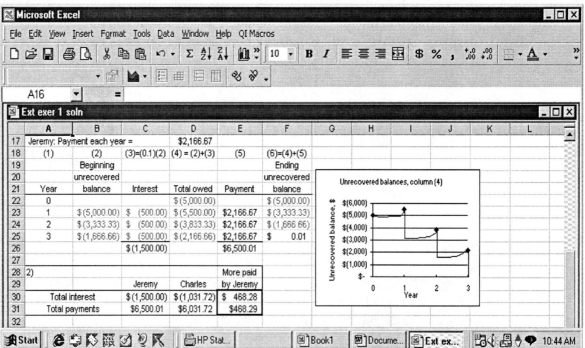

Extended Exercise 2 Solution

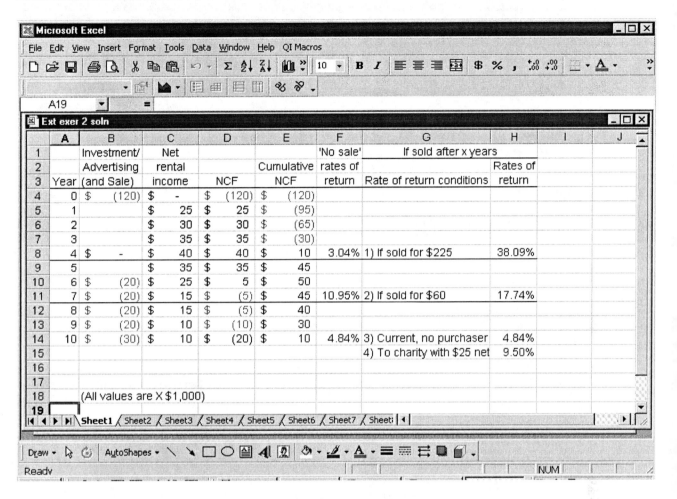

The spreadsheet above summarizes all answers in $1000. Some cells must be changed to obtain the rate of return values shown in column H. These are described below.

1. Use IRR function in year 4 and add $225 in cell C8 for year 4.
    i* (sell after 4) = 38.09%

   With no sale, IRR results in:
    i* (no sale after 4) = 3.04%

2. Use IRR function with $60 added into cell C11 for year 7.
    i* (sell after 7) = 17.74%
    i* (no sale after 7) = 10.95

3.   i* (no sale after 10) = 4.84%

4. Use IRR function with $25 added into cell C14 for year 10.
    i* (charity after 10) = 9.5%

Case Study Solution

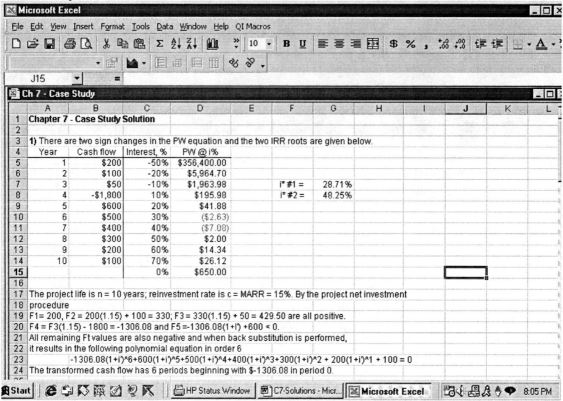

Chapter 7 - Case Study Solution

1) There are two sign changes in the PW equation and the two IRR roots are given below.

Year	Cash flow	Interest, %	PW @ i%			
1	$200	-50%	$356,400.00			
2	$100	-20%	$5,964.70			
3	$50	-10%	$1,963.98		i* #1 =	28.71%
4	-$1,800	10%	$195.98		i* #2 =	48.25%
5	$600	20%	$41.88			
6	$500	30%	($2.63)			
7	$400	40%	($7.08)			
8	$300	50%	$2.00			
9	$200	60%	$14.34			
10	$100	70%	$26.12			
		0%	$650.00			

The project life is n = 10 years; reinvestment rate is c = MARR = 15%. By the project net investment procedure

F1= 200, F2 = 200(1.15) + 100 = 330; F3 = 330(1.15) + 50 = 429.50 are all positive.
F4 = F3(1.15) - 1800 = -1306.08 and F5 =-1306.08(1+i') +600 < 0.
All remaining Ft values are also negative and when back substitution is performed, it results in the following polynomial equation in order 6

$$-1306.08(1+i')^6 + 600(1+i')^5 + 500(1+i')^4 + 400(1+i')^3 + 300(1+i')^2 + 200(1+i')^1 + 100 = 0$$

The transformed cash flow has 6 periods beginning with $-1306.08 in period 0.

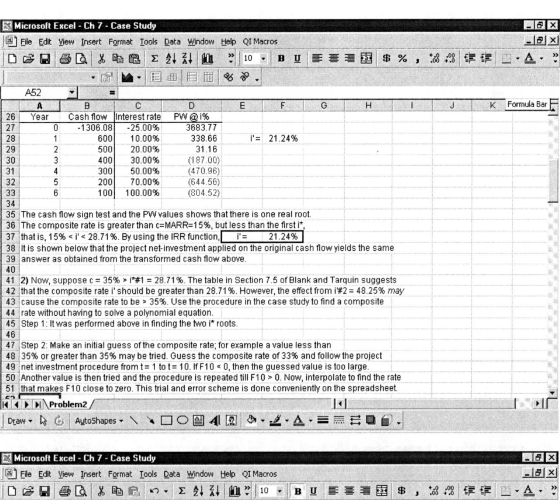

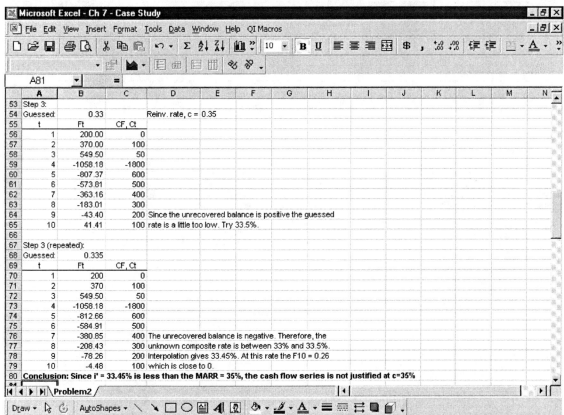

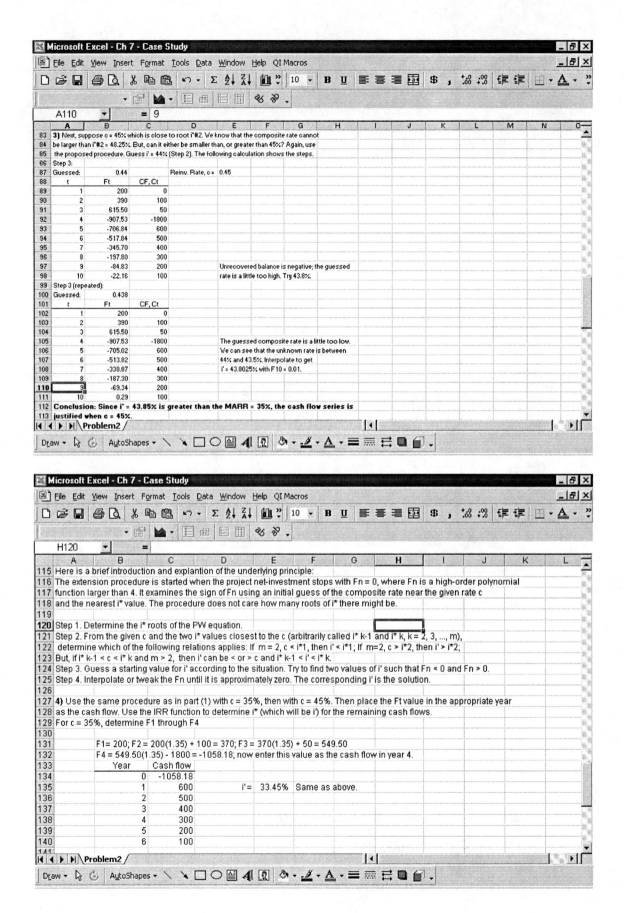

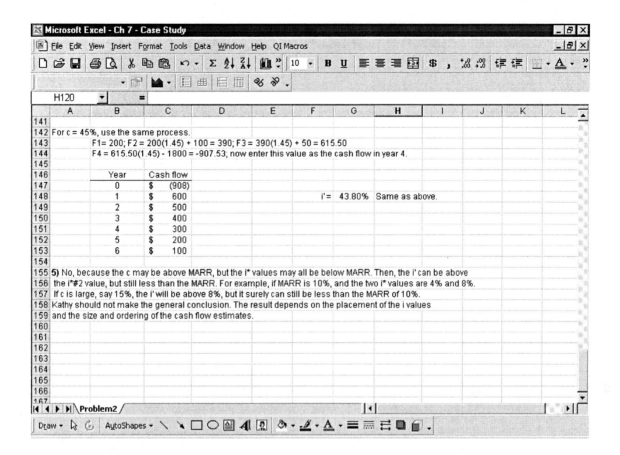

Chapter 8 – Rate of Return Analysis:
Multiple Alternatives
Solutions to end of chapter exercises

Problems

8.1 The rate of return on the increment is higher than the rate of return of both alternatives.

8.2 The microwave model should be purchased because it is the only one with a ROR that exceeds the MARR of 20% per year.

8.3 Scenario (d) is the only one that requires an incremental rate of return analysis because it is not known whether the rate of return on the increment between X and Y is greater than or less than the MARR. For (a), the rate of return on the increment must be greater than 24%, so Y is selected. For (b), only Y has a rate of return above the MARR. For (c), neither alternative has a rate of return above the MARR. For (e), the increment between X and Y has to have a rate of return below 20% per year.

8.4 Overall ROR = [10,000(30%) + 20,000(20%)]/30,000
   = 10% + 13.3%
   = 23.3%

8.5 50,000(0.30) = 10,000(0.15) + 40,000(ROR_{X2})
   ROR_{X2} = 13,500/40,000
   = 33.75%

8.6

	Net cash flows		Incremental
Year	Process A	Process B	B – A
0	$-100,000	$-165,000	$-65,000
1	-60,000	-55,000	+5,000
2	-60,000	-55,000	+5,000
3	-160,000	-55,000	+105,000
4	-60,000	-55,000	+5,000
5	-60,000	-55,000	+5,000
6	-60,000	-55,000	+5,000

8.7

	Net cash flows		Incremental
Month	Alt. F	Alt. G	G – F
0	$-90,000	$-200,000	$-110,000
1 – 24	-12,000	-11,000	+1,000
24	-90,000 / +20,000	0	+70,000
25 – 48	-12,000	-11,000	+1,000
48	+20,000	+30,000	+10,000

8.8 The rate of return on the incremental cash flow column represents the rate of return on the extra investment required for the alternative with the larger initial investment.

8.9 Select the alternative with the lower initial investment.

8.10 (a) For independent projects, no incremental analysis is necessary. Each project's ROR is compared to the do-nothing alternative.

(b) For mutually exclusive alternatives, incremental ROR analysis on the extra investment is necessary. Only the lower investment alternative is compared to the do-nothing alternative.

8.11 These are revenue projects. Remove the first one; include the last two with i* > MARR.

8.12 The higher initial investment is selected. The one with the lower initial investment is preferred at all MARR values above the breakeven rate of return for the incremental cash flow.

8.13 (a) $0 = -3000 + 200(P/A, \Delta i, 3) + (19{,}200 - 14{,}500)(P/F, \Delta i, 3)$

Solve for Δi^* by trial and error or spreadsheet.
$\Delta i^* = 21.95\%$ (RATE or IRR function)

(b) Since $\Delta i^* >$ MARR, the firm should purchase the Toyota.

8.14 $0 = -300{,}000 + 50{,}000(P/A, \Delta i, 5) + (250{,}000 - 100{,}000)(P/F, \Delta i, 5)$

Solve for Δi^* by trial and error or spreadsheet.
$\Delta i^* = 8.17\%$ (RATE or IRR function)

Select process X.

8.15 The rate of return equation for the incremental cash flow of (B–A) is:

$0 = -10{,}000 + 1200(P/A, \Delta i, 6) + (15{,}000 - 3000)(P/F, \Delta i, 3) + 3000(P/F, \Delta i, 6)$

Solve for Δi^* by trial and error or spreadsheet.
$\Delta i^* = 26.8\%$ (IRR function)

Select machine B.

8.16 The rate of return equation for the incremental cash flow of (Q − P) is:

$$0 = -17{,}000 + 400(P/A, \Delta i, 9) + (18{,}000 - 3000)(P/F, \Delta i, 3) + (18{,}000 - 3000)(P/F, \Delta i, 6) - 300(P/F, \Delta i, 9)$$

Solve for Δi^* by trial and error or spreadsheet.
$\Delta i^* = 16.9\%$ (IRR function)

Select alternative P.

8.17 $0 = -90{,}000 + 10{,}000(P/A, \Delta i, 5) + 20{,}000(P/A, \Delta i, 4)(P/F, \Delta i, 5) + 5000(P/F, \Delta i, 10)$

Solve for Δi^* by trial and error or spreadsheet.
$\Delta i^* = 7.4\%$ (IRR function)

Select alternative J.

8.18 The rate of return on the incremental cash flow is 46.6%. The first cost of S that will result in a rate of return of exactly 50% is:

$P = 400{,}000(P/F, 50\%, 1) + 700{,}000(P/F, 50\%, 2) + 950{,}000(P/F, 50\%, 3)$
$ = 400{,}000(0.6667) + 700{,}000(0.4444) + 950{,}000(0.2963)$
$ = \$859{,}245$

The required reduction in the first cost of S is:
Reduction = $900,000 − 859,245
$ = \$40{,}755$

8.19 (a) The breakeven rate of return is the Δi^* value at which the equivalent costs (PW, AW, FW) of the alternatives are the same. Equate the PW values and solve for Δi^*. The difference equation is actually the rate of return equation for the incremental cash flow.

$-100{,}000 - 50{,}000(P/A, \Delta i, 5) + 20{,}000(P/F, \Delta i, 5)$
$\quad = -175{,}000 - 30{,}000(P/A, \Delta i, 5) + 40{,}000(P/F, \Delta i, 5)$

$0 = -75{,}000 + 20{,}000(P/A, \Delta i, 5) + 20{,}000(P/F, \Delta i, 5)$

Solve for Δi^* by trial and error or spreadsheet.
$\Delta i^* = 16.0\%$ (RATE or IRR function)

(b) The spreadsheet below is a plot of PW versus i for the incremental cash flows. All values of MARR above $\Delta i^* = 16.0\%$ require that R be selected.

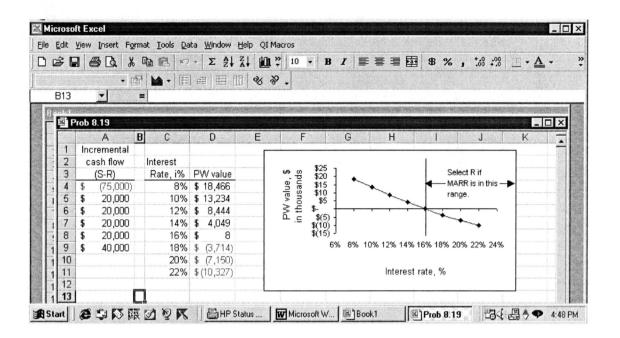

8.20 $0 = -25{,}000(A/P, \triangle i, 10) + 5000 + 3000(A/F, \triangle i, 10)$

Solve for $\triangle i^*$ by trial and error or spreadsheet.
$\triangle i^* = 15.8\%$ (RATE or IRR function)

Select alternative M.

8.21 (a) $0 = -50{,}000(A/P, \triangle i, 6) + 15{,}000 + (40{,}000 - 5000)(P/F, \triangle i, 3)(A/P, \triangle i, 6)$
$+ (11{,}000 - 5000)(A/F, \triangle i, 6)$

Solve for $\triangle i^*$ by trial and error or spreadsheet.
$\triangle i^* = 35.95\%$ (IRR function)

Select the automatic machine.

(b) The spreadsheet plots AW versus $\triangle i$ for the incremental cash flows and shows the breakeven point just above 35%.

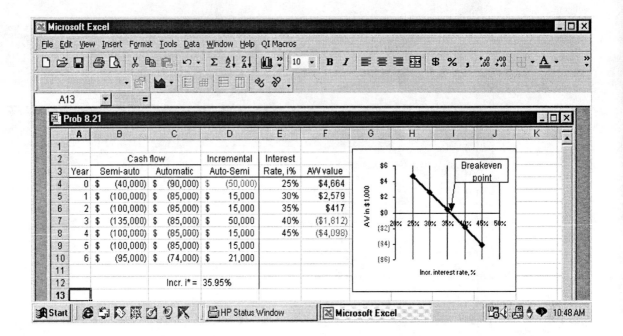

8.22 $0 = -50{,}000(A/P, \triangle i, 0) + 10{,}000 + 1000(A/G, \triangle i, 10)$

Solve for $\triangle i^*$ by trial and error or spreadsheet.
  $\triangle i^* = 22.5\%$ (IRR function)

Select alternative Z1.

8.23 Since the bridge will have an infinite life, the incremental rate of return equation is:

$0 = \$-500{,}000(\triangle i) + 20{,}000$

$\triangle i = 20{,}000/500{,}000$

$\triangle i^* = 4\%$ per year

Select design 1A, since $4\% <$ MARR $= 10\%$.

8.24 Since the rate of return for each increment of investment exceeded the MARR, the alternative that required the highest initial investment should be selected.

8.25 Revenue alternatives have income amounts as part of their cash flow estimates while service alternatives do not. Service alternatives assume revenues are the same for all alternatives.

8.26 An alternative's rate of return is the same as an incremental rate of return between the alternative and the do-nothing alternative.

Chapter 8

8.27 When the rate of return on all increments of investment for service alternatives is less than the MARR, the alternative that requires the lowest initial investment should be selected.

8.28 (a) The ROR for each method is compared to MARR. Use the RATE or IRR function for each cash flow series.

    A: $i^* = 23.67\%$
    B: $i^* = 25.12\%$
    C: $i^* = 20.09\%$
    D: $i^* = 18.66\%$

All methods are economically acceptable if they are independent projects.

(b) The alternatives are already ranked in terms of increasing initial investment. First, compare Method A against do-nothing (DN).

<u>A to DN</u>

$0 = -15{,}000 + 4000(P/A, \Delta i, 10) + 1000(P/F, \Delta i, 10)$

    $\Delta i^* = 23.67\%$ (RATE or IRR function)

Since $i^* > 12\%$, method A is acceptable; eliminate DN.

<u>B to A</u>

$0 = -3000 + 1000(P/A, \Delta i, 10) + 1000(P/F, \Delta i, 10)$

    $\Delta i^* = 31.91\% > 12\%$, eliminate A.

<u>C to B</u>

$0 = -7000 + 1000(P/A, \Delta i, 10) - 2500(P/F, \Delta i, 10)$

    $\Delta i^* = 1.75\% < 12\%$, eliminate C.

<u>D to B</u>

$0 = -17{,}000 + 3000(P/A, \Delta i, 10) - 2700(P/F, \Delta i, 10)$

    $i_\Delta^* = 10.57\% < 12\%$, eliminate D.

Therefore, select method B.

8.29 Rank the alternatives by increasing initial investment: 1, 3, 4, 5, 2

 3 to 1

 $0 = -4000 + 1000(P/A, \Delta i, 10)$
  $(P/A, \Delta i, 10) = 4.0000$
  $\Delta i^* = 21.4\% < 25\%$, eliminate 3.

 4 to 1

 $0 = -5000 + 2000(P/A, \Delta i, 10)$
  $(P/A, \Delta i, 10) = 2.5000$
  $\Delta i^* = 38.5\% > 25\%$, eliminate 1.

 5 to 4

 $0 = -13,000 + 4000(P/A, \Delta i, 10)$
  $(P/A, \Delta i, 10) = 3.2500$
  $\Delta i^* = 28.2\% > 25\%$, eliminate 4.

 2 to 5

 $0 = -5000 + 1500(P/A, \Delta i, 10)$
  $(P/A, \Delta i, 10) = 3.3333$
  $\Delta i^* = 27.3\% > 25\%$, eliminate 5.

 Therefore, select machine 2.

8.30 (a) The alternatives are already ranked according to increasing initial investment cost. Therefore, compare 8 with do-nothing (DN) first.

 8 to DN

 $0 = -10,000 + (6500 - 4000)(P/A, \Delta i, 8) + 2000(P/F, \Delta i, 8)$
  $\Delta i^* = 20.2\%$, eliminate DN.

 10 to 8

 $0 = -4,000 + 2000(P/A, \Delta i, 8) + 500(P/F, \Delta i, 8)$
  $\Delta i^* = 48.1\%$, eliminate 8.

 15 to 10

 $0 = -4000 + (4000 - 1500)(P/A, \Delta i, 8) + 500(P/F, \Delta i, 8)$
  $\Delta i^* = 61.3\%$, eliminate 10.

20 to 15

$0 = -6000 + 2500(P/A, \triangle i, 8) + 500(P/F, \triangle i, 8)$
$\triangle i^* = 38.9\%$, eliminate 15.

25 to 20

$0 = -9000 + 1000(P/A, \triangle i, 8) + 3100(P/F, \triangle i, 8)$
$\triangle i^* = 4\%$, eliminate 25.

Select size 20 cubic meters.

(b) Buy another 20 cubic meter truck. Always spend the most that is justified, that is, that has an extra investment with $\triangle i^* >$ MARR.

8.31 (a) If projects are independent, select all that have a rate of return > 13%. Therefore, select A, B, C, D.

(b) Must conduct incremental analysis. Rank projects in terms of increasing initial investment cost: B, C, A, E, D. E can be eliminated since there are revenue projects and $i_E^* = 12\% <$ MARR. Evaluate the rest.

B is justified because its ROR > 13%. Compare B and C.

C to B

$\triangle i^* = \dfrac{900}{5000}$
$= 18\% > 13\%$, eliminate B.

A to C

$\triangle i^* = \dfrac{1100}{5000}$
$= 22\% > 13\%$, eliminate C.

D to A

$\triangle i^* = \dfrac{6000}{50,000}$
$= 12\% < 13\%$, eliminate D.

Select project A.

(c) Project D is selected. It is incorrect because the $50,000 investment increment over A returns only 12%, which is less than MARR = 13%. Incremental analysis is necessary to select A.

8.32 (a) Select machine 2, because it is the only one with ROR > 20%.

(b) Machine 1 is acceptable because its ROR > 18%. Therefore, compare 2-to-1 incrementally. Eliminate 1 because incremental ROR of 35.7% > 18%. Compare 3-to-2 incrementally and eliminate 3 because –3.2% < 18%. Machine 4 is not acceptable because its ROR < 18% and all incremental $ values are < 0. Therefore, select machine 2.

8.33 (a) Only alternative A has a ROR > 18%.

(b) <u>A to DN</u>
  $\triangle i^* = 29\% > 14.5\%$, eliminate DN,

  <u>B to A</u>
  $\triangle i^* = 1\% < 14.5\%$, eliminate B.

  <u>C to A</u>
  $\triangle i^* = 7\% < 14.5\%$, eliminate C.

  <u>D to A</u>
  $\triangle i^* = 10\% < 14.5\%$, eliminate D.

  Therefore, select alternative A.

(c) <u>A to DN</u>
  $\triangle i^* = 29\% > 9.5\%$, eliminate DN.

  <u>B to A</u>
  $\triangle i^* = 1\% < 9.5\%$, eliminate B.

  <u>C to A</u>
  $\triangle i^* = 7\% < 9.5\%$, eliminate C.

  <u>D to A</u>
  $\triangle i^* = 10\% > 9.5\%$, eliminate A.

  Therefore, select alternative D.

8.34 (a) Find the annual income for each alternative using A = P×(ROR).

$$\text{Income}_E = 20,000(0.30)$$
$$= \$6000$$

$$\text{Income}_F = 35{,}000(0.25)$$
$$= \$8{,}750$$
$$\text{Income}_G = 50{,}000(0.19)$$
$$= \$9{,}500$$
$$\text{Income}_H = 90{,}000(0.21)$$
$$= \$18{,}900$$

Incremental ROR for F to E:

$$\Delta i^* = \frac{(8750 - 6000)}{(35{,}000 - 20{,}000)} = 18.3\%$$

Incremental ROR for G to E

$$\Delta i^* = \frac{3500}{30{,}000} = 11.7\%$$

Incremental ROR for H to E

$$\Delta i^* = \frac{12{,}900}{70{,}000} = 18.4\%$$

Incremental ROR for G to F

$$\Delta i^* = \frac{750}{15{,}000} = 5\%$$

Incremental ROR for H to F

$$\Delta i^* = \frac{10{,}150}{55{,}000} = 18.5\%$$

Incremental ROR for H to G

$$\Delta i^* = \frac{9400}{40{,}000} = 23.5\%$$

(b) E to DN: $\Delta i^* = 30\%$ eliminate DN

F to E: $\Delta i^* = 18.3\%$ eliminate E

G to F: $\Delta i^* = 5\%$ eliminate G

H to F: $\Delta i^* = 18.5\%$ eliminate F

Select H.

(c) Perform the incremental ROR analysis with MARR = 19%.

    E to DN: $\Delta i^* = 30\%$ eliminate DN

    F to E: $\Delta i^* = 18.3\%$ eliminate F

    G to E: $\Delta i^* = 11.7\%$ eliminate G

    H to E: $\Delta i^* = 18.4\%$ eliminate H

Select E for the first alternative. For the second best alternative, perform the same comparison, but don't include E.

    F to DN: $\Delta i^* = 25\%$ eliminate DN

    G to F: $\Delta i^* = 5\%$ eliminate G

    H to F: $\Delta i^* = 18.5\%$ eliminate H

Select F, as the second best alternative.

FE Review Solutions

8.35 Since the rate of return *decreased* to 20% as a result of the extra investment, the rate of return on the increment has to be less than 20%.

Answer is (d).

8.36 The rate of return on the increment has to be less than 16% because the alternative's ROR decreased with the increased investment. Therefore, alternative X is the better choice.

Answer is (a).

8.37 Answer is (d).

8.38 Answer is (d).

8.39 Answer is (d).

8.40 Answer is (c).

8.41 Answer is (b).

8.42 Answer is (d)

Extended Exercise Solution

1. PW at 12% is shown in row 29. Select #2 (n = 8) with the largest PW value.

2. #1 (n = 3) is eliminated. It has i* < MARR = 12%. Perform an incremental analysis of #1 (n = 4) and #2 (n = 5). Column H shows △i* = 19.49%. Now perform an incremental comparison of #2 for n = 5 and n = 8. This is not necessary. No extra investment is necessary to expand cash flow by three years. The △i* is infinity. It is obvious: select #2 (n = 8).

3. PW at 2000% > $0.05. △i* is infinity, as shown in cell K45, where an error for IRR(K4:K44) is indicated. This analysis is not necessary, but shows how Excel can be used over the LCM to find a rate of return.

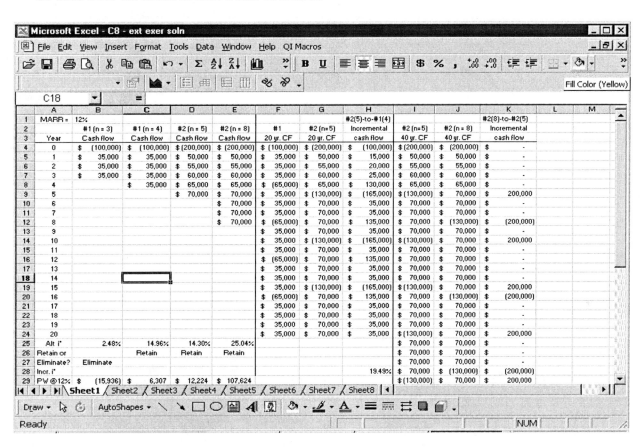

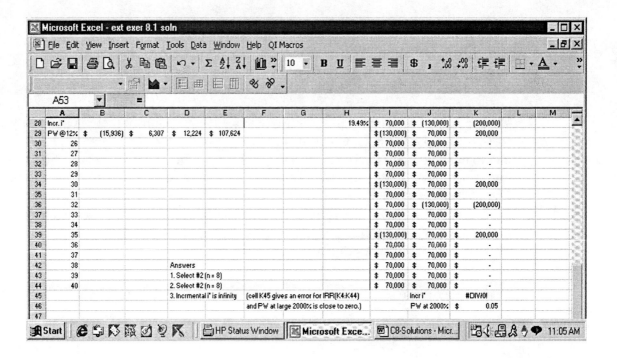

Case Study 1 Solution

1. Cash flows for each option are summarized at top of the spreadsheet. Rows 9-19 show annual estimates for options in increasing order of initial investment: 3, 2, 1, 4, 5.

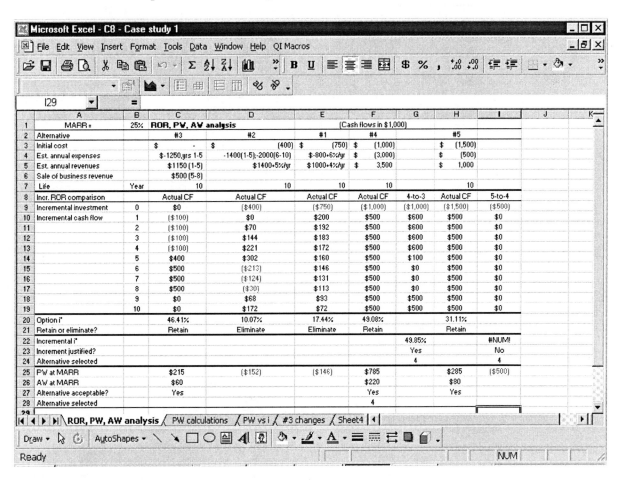

2. Multiple i* values: Only for option #2; there are 3 sign changes in cash flow and cumulative cash flow series. No values other than 10.07% are found in the 0 to 100% range.

3. Do incremental ROR analysis after removing #1 and #2. See row 22. 4-to-3 comparison 4-to-3 yields 59.85%, 5-to-4 has no return because all incremental cash flows are 0 or negative. PW at 25% is $785.25 for #4, which is the largest PW.

   Conclusion: Select option #4 – trade-out with friend.

4. PW vs. i charts for all 5 options are on the spreadsheet.

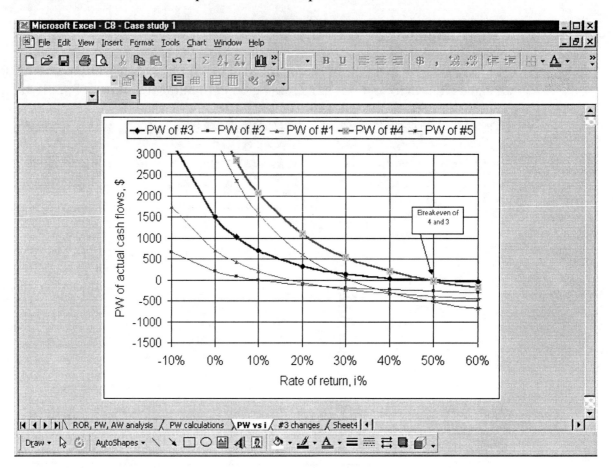

Options Compared	Approximate breakeven
1 and 2	26%
3 and 5	27
2 and 5	38
1 and 5	42
3 and 4	50

5. Force the breakeven rate of return between options #4 and #3 to be equal to MARR = 25%. Use trial and error or Solver on Excel with a target cell of G22 (to be 25% or .25 on the Solver window) and changing cell of C6. Make the values in years 5 through 8 of option #3 equal to the value in cell C6, so they reflect the changes. The answer obtained should be about $1090, which is actually $1,090,000 for each of 4 years.

Required minimum selling price is 4(1090,000) = $4.36 million.

Chapter 8

Case Study 2 Solution

1. By inspection only: Select Plan A since its cash flow total at 0% is $300, while Plan B produces a loss of the same amount ($-300).

2. Calculations for the following are shown on the spreadsheet below.

    PW at MARR approach:

    PW at 15%: $PW_A = -\$81.38$ and $PW_B = +\$81.38$. Plan B is selected
    PW at 50%: $PW_A = +\$16.05$ and $PW_B = -\$16.05$. Plan A is selected.

    The decisions contradict each other when the MARR is different. It does not seem logical to accept plan A at a higher interest rate (50%) and at 0%, but reject it at a mid-point interest rate (15%). The PW at MARR method is not working!

    ROR approach:

    The cash flow series have two sign changes, so a maximum of two roots may be found. An ROR analysis using Excel functions for the two plans produce two identical roots for each plan:

    $$i^*_1 = 9.51\% \quad \text{and} \quad i^*_2 = 48.19\%$$

    There are two i* values; it is not clear which value to use for a decision on project acceptability. Further, when there are multiple i* values, the PW analysis 'at the MARR' does not work, as demonstrated above.

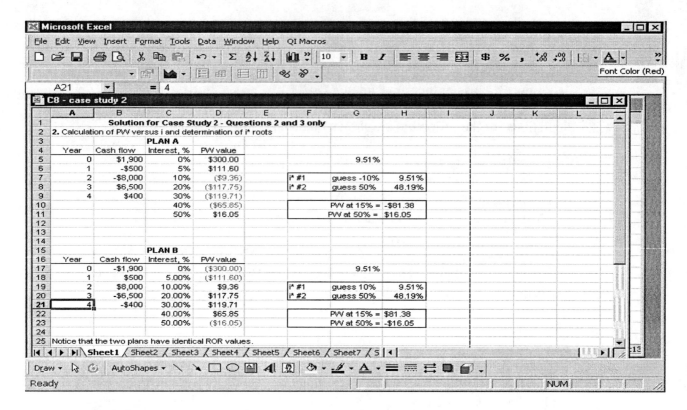

3. Incremental ROR analysis is shown on the spreadsheet below.

> Plan B has a larger initial investment than A. The incremental cash flow series (B-A) has two sign changes. The use of the IRR function finds the same two roots: 9.51% and 48.19%. Incremental ROR analysis offers no definitive resolution.

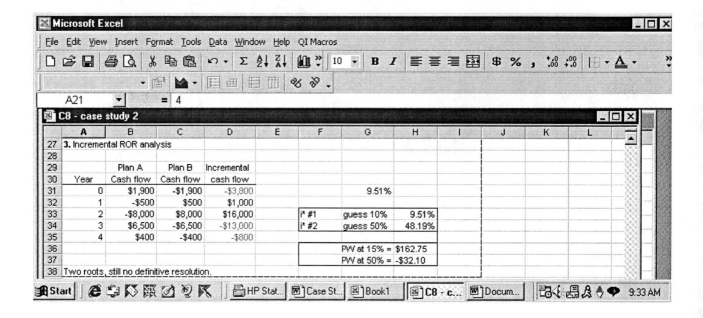

Chapter 8 - 17 -

4. Composite rate of return approach

Plan A (a) MARR = 15% and c = 15%

$F_0 = 1900$; $F_1 = 1900(1.15) - 500 = 1685$; $F_2 = 1685(1.15) - 8000 = -6062.25$;
$F_3 = -6062.25(1+i') + 6500 = 437.75 - 6062.25i'$
$F_4 = 0 = F_3(1+i') + 400$
 So $i' = 13.06\% <$ MARR = 15%. Reject Plan A.

(b) MARR = 15% and c = 45%

$F_0 = 1900$; $F_1 = 1900(1.45) - 500 = 2255$;
$F_2 = 2255(1.45) - 8000 = -4730.25$; $F_3 = -4730.25(1+i') + 6500 = 1769.75 - 4730.25i'$ $F_4 = 0 = F_3(1+i') + 400$.
 So $i' = 43.31\% >$ MARR = 15%. Accept Plan A.

(c) MARR = 50% and c = 50%

$F_0 = 1900$; $F_1 = 1900(1.50) - 500 = 2350$; $F_2 = 2350(1.50) - 8000 = -4475$;
$F_3 = -4475(1+i') + 6500 = -4475i' + 2025$
$F_4 = 0 = F_3(1+i') + 400$.
 So $i' = 51.16\% >$ MARR = 50%. Accept Plan A.

Plan B (a) MARR = 15% and c = 15%

$F_0 = -1900$; $F_1 = -1900(1+i') + 500$; $F_2 = -1900(1+i')^2 + 500(1+i') + 8000$;
$F_3 = (1.15)F_2 - 6500$
$F_4 = 0 = F_3(1.15) - 400$.
 So $i' = 17.74\% >$ MARR = 15%. Accept Plan B.

(b) MARR = 15% and c = 45%

$F_0 = -1900$; $F_1 = -1900(1+i') + 500$; $F_2 = -1900(1+i')^2 + 500(1+i') + 8000$;
$F_3 = -1900(1.45)(1+i')^2 + 500(1.45)(1+i') + 8000(1.45) - 6500$
$F_4 = 0 = F_3(1.45) - 400$.
 So $i' = 46.14\% >$ MARR = 15%. Accept Plan B

(c) MARR = 50% and c = 50%

$F_0 = -1900$; $F_1 = -1900(1+i') + 500$; $F_2 = -1900(1+i')^2 + 500(1+i') + 8000$;
$F_3 = -1900(1.5)(1+i')^2 + 500(1.5)(1+i') + 8000(1.5) - 6500$
$F_4 = 0 = F_3(1.5) - 400$.
 So $i' = 49.30\% <$ MARR = 50%. Reject Plan B.

d) Discussion: Plan B is superior to Plan A for c values below i^*_2, i.e., B's composite rate of return is higher. However, for c values above i^*_2, plan A gives a higher (composite) rate of return.

Conclusion: The composite rate of return evaluation yields unambiguous results when a reinvestment rate is specified.

Chapter 9 – Benefit/Cost Analysis and Public Sector Economics
Solutions to end of chapter exercises

Problems

9.1 (a) Public sector projects usually require large initial investments while many private sector investments may be medium to small.

(b) Public sector projects usually have long lives (30-50 years) while private sector projects are usually in the 2-25 year range.

(c) Public sector projects are usually funded from taxes, government bonds, or user fees. Private sector projects are usually funded by stocks, corporate bonds, or bank loans.

(d) Public sector projects use the term discount rate, not MARR. The discount rate is usually in the 4 – 10% range, thus it is lower than most private sector MARR values.

9.2 (a) Private (b) Private (c) Public (d) Public (e) Public
(f) Private

9.3 (a) Cost (b) Benefit (c) Cost (d) Disbenefit (e) Benefit

9.4 (a) B/C = $\frac{600,000 - 90,000}{550,000}$

= 0.93

(b) B-C = 600,000 – 90,000 – 550,000
= $-40,000

9.5 $1.0 = \frac{10 * \text{value of 1 life}}{200 * 80,000,000}$

Value = $1.6 billion per human life

9.6 Annual cost = 30,000 * 0.025
= $750 per year/household
Let x = number of households

Total annual cost, C = (750)(x)

Total annual benefits, B = (0.01)(x)($/household)

$$1.0 = \frac{B}{C} = \frac{B}{(750)(x)}$$

$$B = (750)(x)$$

Substitute B = (0.01)(x)($/household):

(0.01)(x)($/household) = 750x

$/household = $75,000 per year

9.7 (a) Cost = 2,000,000(0.08) + 100,000
 = $260,000 per year

$$B/C = \frac{330,000 - 40,000}{260,000}$$

 = 1.12

(b) Cost = 2,000,000(0.08)
 = $160,000 per year

$$\text{Modified B/C} = \frac{330,000 - 40,000 - 100,000}{160,000}$$

 = 1.19

9.8 Cost = 150,000(A/P,6%,20) + 12,000
 = 150,000(0.08718) + 12,000
 = $25,077 per year

Benefits = 20,000(2)(0.50)
 = $20,000 per year

$$B/C = \frac{20,000}{25,077}$$

 = 0.80

9.9 $$1.3 = \frac{500,000}{P(0.07) + 200,000}$$

P = $2,637,362

9.10 $$1.7 = \frac{150,000 - M\&O}{1,000,000(A/P,6\%,30)}$$

$$1.7 = \frac{150,000 - M\&O}{1,000,000(0.07265)}$$

M&O = $26,495 per year

9.11 $2.1 = \dfrac{400{,}000 - 25{,}000}{P(A/P,8\%,10) + 150{,}000}$

  $2.1 = \dfrac{375{,}000}{P(0.14903) + 150{,}000}$

  P = $191,716

9.12 Convert all estimates to PW values as necessary.

  PW disbenefits = 45,000(P/A,10%,20)
    = 45,000(8.5136)
    = $383,112
  PW M&O Cost = 300,000(P/A,10%,20)
    = 300,000(8.5136)
    = $2,554,080
  B/C = $\dfrac{3{,}800{,}000 - 383{,}112}{1{,}200{,}000 + 2{,}554{,}080}$
   = 0.91

9.13 Cost = 30,000,000(0.08) + 100,000
    = $2,500,000 per year

  B/C = $\dfrac{2{,}800{,}000}{2{,}500{,}000}$
   = 1.12

  Construct the dam.

9.14 AW = C = 2,200,000(0.12) + 10,000 + 65,000(A/F,12%,15)
    = 264,000 + 10,000 + 65,000(0.02682)
    = $275,743
  Annual Benefit = B = 90,000 − 10,000
    = $80,000
  B/C = $\dfrac{80{,}000}{275{,}743}$
   = 0.29

  Since B/C < 1.0, the dam should <u>not</u> be constructed.

9.15 AW = C = 6,000,000(A/P,6%,25) + 20,000
    = 6,000,000(0.07823) + 20,000
    = $489,380

  Annual Benefit = B = $350,000

(a) B − C = $350,000 − 489,380
       = $-139,380
   Since (B − C) < 0, do not construct the highway.

(b) B/C = 350,000/489,380
       = 0.715
   Since B/C < 1.0, do not construct the highway.

(c) For modified B/C ratio, $20,000 maintenance is subtracted from benefits.
   Modified B/C = (350,000 − 20,000)/6,000,000(0.07823)
              = 0.70
   Since B/C < 1, do not construct the highway.

9.16 (a) AW = C = 1,500,000 (A/P, 6%, 20) + 25,000
               = 1,500,000 (0.08718) + 25,000
               = $155,770

   Annual revenue = B = $175,000

   B/C = 175,000/155,770
       = 1.12
   Since B/C > 1.0, the canals should be extended.

(b) For modified B/C ratio,
   C = 1,500,000(A/P,6%,20) = $130,770
   B = 175,000 − 25,000 = 150,000

   Modified B/C = 150,000/130,770
               = 1.15
   Since B/C > 1, canals should be extended.

9.17   AW = C = 1,500,000(A/P,6%,20) + 25,000 + 60,000[(P/F,6%,3) + (P/F,6%,6)
           + (P/F,6%,9) + (P/F,6%,12) + (P/F,6%,15) + (P/F,6%,18)](A/P,6%,20)
         = 1,500,000 (0.08718) + 25,000 + 60,000 [0.8396 + 0.7050 + 0.5919 +
           0.4970 + 0.4173 + 0.3503](0.08718)
         = $173,560

Annual disbenefit = D = $15,000
Annual revenue = B = $175,000
(B − D)/C = (175,000 − 15,000)/173,560
         = 0.922

The disbenefit of $15,000 per year and the dredging cost each third year have reduced the B/C ratio to below 1.0; the canals should not be extended now.

9.18 Annual cost of long route = 21,000,000(0.06) + 40,000 + 21,000,000(0.10)(A/F,6%,10)
= 1,260,000 + 40,000 + 2,100,000(0.07587)
= $1,459,327

Annual cost of short route = 45,000,000(0.06) + 15,000 + 45,000,000(0.10)(A/F,6%,10)
= 2,700,000 + 15,000 + 4,500,000(0.07587)
= $3,056,415

The short route must be incrementally justified.

Extra cost for short route = 3,056,415 − 1,459,327
= $1,597,088

Incremental benefits of short route = 400,000(0.35)(25 − 10) + 900,000
= $3,000,000

$$\triangle B/C_{short} = \frac{3,000,000}{1,597,088} = 1.88$$

Build the short route.

9.19 Justify extra cost of downtown location.
Extra cost for DT site = 11,000,000(0.08)
= $880,000
Extra benefits for DT site = 350,000 + 400,000
= $750,000
Incremental B/C_{DT} = 750,000/880,000
= 0.85

The city should build on the west-side site.

9.20 First compare program 1 to do-nothing (DN).

Cost/household/mo = $60(A/P,0.5%,60)
= 60(0.01933)
= $1.16

$$B/C_1 = \frac{1.25}{1.16} = 1.08 \quad \text{Eliminate DN.}$$

Compare program 2 to program 1.

$$\Delta\text{cost} = 500(A/P, 0.5\%, 60) - 60(A/P, 0.5\%, 60)$$
$$= (500 - 60)(0.01933)$$
$$= \$8.51$$
$$\Delta\text{benefits} = 8 - 1.25$$
$$= \$6.75$$
$$\Delta B/C_2 = \frac{6.75}{8.51}$$
$$= 0.79 \quad \text{Eliminate program 2.}$$

The utility should undertake program 1.

9.21 Alternative 2 has a larger total cost; it must be incrementally justified. Use PW values. Benefit is the difference in damage costs.

For alternative 2:

$$\Delta\text{cost} = (1{,}100{,}000 - 600{,}000) + (70{,}000 - 50{,}000)(P/A, 8\%, 20)$$
$$= \$500{,}000 + 20{,}000(9.8181)$$
$$= \$696{,}362$$
$$\Delta\text{benefit} = (950{,}000 - 250{,}000)(P/F, 8\%, 10)$$
$$= 700{,}000(0.4632)$$
$$= 324{,}240$$
$$\Delta B/C = \frac{324{,}240}{696{,}362}$$
$$= 0.47$$

Select alternative 1.

9.22 Using the capital recovery costs, solar is the more costly alternative.

$$\Delta\text{cost} = (1{,}500{,}000 - 200{,}000)(A/P, 0.5\%, 36) - (150{,}000 - 40{,}000)(A/F, 0.5\%, 36)$$
$$= 1{,}300{,}000(0.03042) - 110{,}000(0.02542)$$
$$= \$36{,}750$$
$$\Delta\text{benefits} = 50{,}000 - 3{,}000$$
$$= \$47{,}000$$
$$\Delta B/C = \frac{47{,}000}{36{,}750}$$
$$= 1.28$$

Select the solar system.

9.23 (a) Location E
AW = C = 3,000,000(0.12) + 50,000
= $410,000
Revenue = B = $500,000 per year
Disbenenfits = D = $30,000 per year

Location W
AW = C = 7,000,000 (0.12) + 65,000 - 25,000
= $880,000
Revenue = B = $700,000 per year
Disbenefits = D = $40,000 per year

B/C ratio for location E:
(B – D)/C = (500,000 – 30,000)/410,000
= 1.15

Location E is economically justified. Location W is now incrementally compared to E.

$\triangle$cost of W = 880,000 – 410,000
= $470,000
$\triangle$benefits of W = 700,000 – 500,000
= $200,000
$\triangle$disbenefits of W = 40,000 – 30,000
= $10,000
$\triangle$B/C = (B – D)/C = (200,000 – 10,000)/470,000
= 0.40

Since (B – D)/C < 1, W is not justified. Select location E.

(b) Location E
B = 500,000 – 30,000 – 50,000 = $420,000
C = 3,000,000 (0.12) = $360,000
Modified B/C = 420,000/360,000
= 1.17

Location E is justified.

Location W
$\triangle$B = $200,000
$\triangle$D = $10,000
$\triangle$C = (7 million – 3 million)(0.12)
= $480,000
$\triangle$M&O = (65,000 – 25,000) – 50,000
= $-10,000
Note that M&O is now an incremental cost advantage for W.

Chapter 9

Modified B/C = $\dfrac{200,000 - 10,000 + 10,000}{480,000}$

= 0.42

W is not justified; select location E.

9.24 <u>Annual worth of costs</u> For the dome, remove $800,000 from first cost and amortize separately over 10 years.

$C = AW_{Dome}$ = 299,200,000(A/P,8%,50) + 300,000 + 10,000(A/G,8%,50)
  + 800,000(A/P,8%,10) − 5,000,000(A/F,8%,50)
= 299,200,000(0.08174) + 300,000 + 10,000(11.4107)
  + 800,000(0.14903) − 5,000,000(0.00174)
= $24,981,239

For the conventional stadium, remove $100,000 from the first cost and amortize separately over 4 years.

$C = AW_{Conv.}$ = 49,900,000(A/P,8%,50) + 75,000 + 8000(A/G,8%,50)
  + 100,000(A/P,8%,4) − 5,000,000(A/F,8%,50)
= 49,900,000 (0.08174) + 75,000 + 8000(11.4107)
  + 100,000(0.30192) − 5,000,000(0.00174)
= $4,266,604

ΔC = 24,981,239 − 4,266,604 = $20,714,635

The dome must be incrementally justified.

Present worth of ΔB = [500,000 + 200,000(A/G,8%,15)](P/A,8%,15)
  + 3,300,000(P/A,8%,35)(P/F,8%,15)
= [500,000 + 200,000(5.5945)](8.5595)
  + 3,300,000(11.6546)(0.3152)
= $25,979,623

AW of ΔB = 25,979,623(A/P,8%,50)
= 25,979,623(0.08174)
= $2,123,574

$\dfrac{\Delta B}{\Delta C} = \dfrac{2,123,574}{20,714,635}$ = 0.10

Construct conventional stadium.

Chapter 9

9.25 Method 1 vs DN
$C = 15,000(A/P,15\%,10)$
$ = 15,000(0.19925)$
$ = \$2,988.75$

If operating cost is subtracted from benefits:
$B = 15,000 - 10,000 = 5000$
$B/C = 5000/2,988.75$
$ = 1.67 > 1.0$ .... Eliminate DN, keep method 1.

Method 2 vs Method 1
$\Delta C = (19,000 - 15,000)(A/P,15\%,10)$
$ = 4000(0.19925)$
$ = \$797$
$\Delta B = (20,000 - 12,000) - (15,000 - 10,000)$
$ = 3000$
$B/C = 3000/797$
$ = 3.76 > 1$ .... Eliminate method 1, keep method 2.

Method 3 vs Method 2
$\Delta C = (25,000 - 19,000)(A/P,15\%,10)$
$ = 6000(0.19925)$
$ = \$1195.50$
$\Delta B = (19,000 - 9000) - (20,000 - 12,000)$
$ = \$2000$
$B/C = 2000/1195.50$
$ = 1.67 > 1.0$ .... Eliminate method 2, keep method 3.

Method 4 vs Method 3
$\Delta C = (33,000 - 25,000)(A/P,15\%,10)$
$ = 8000(0.19925)$
$ = \$1594$
$\Delta B = (22,000 - 11,000) - (19,000 - 9000)$
$ = 1000$
$B/C = 1000/1594$
$ = 0.63 < 1.0$ .... Eliminate method 4.

Use method 3.

9.26 Cannot consider "do-nothing" alternative, since only costs are given. Combine the investment and installation costs.

<u>160 vs 140mm</u>
$\Delta C = (11{,}310 - 9780)(A/P,8\%,15)$
$\quad = 1530(0.11683)$
$\quad = \$178.75$
$\Delta B = 6000 - 5800$
$\quad = \$200$
$B/C = 200/178.75$
$\quad\quad = 1.12 > 1.0 \quad\quad$ Eliminate 140 mm size.

<u>200 vs 160 mm</u>
$\Delta C = (14{,}580 - 11{,}310)(A/P,8\%,15)$
$\quad = 3270(0.11683)$
$\quad = \$382.03$
$\Delta B = 5800 - 5200$
$\quad = \$600$
$B/C = 600/382.03$
$\quad\quad = 1.57 > 1.0 \quad\quad$ Eliminate 160 mm size.

<u>240 vs 200</u>
$\Delta C = (17{,}350 - 14{,}580)(A/P,8\%,15)$
$\quad = 2770(0.11683)$
$\quad = \$323.62$
$\Delta B = 5200 - 4900$
$\quad = \$300$
$B/C = 0.93 < 1.0 \quad\quad$ Eliminate 240 mm size.

Select the 200 mm pipe size.

9.27 <u>Compare A vs DN:</u>
AW of Cost $= 50(A/P,10\%,5) + 3$
$\quad\quad\quad\quad = 50(0.26380) + 3$
$\quad\quad\quad\quad = 16.19$
AW of Benefits $= 20 - 0.5$
$\quad\quad\quad\quad\quad = 19.5$
$B/C = \dfrac{19.5}{16.19}$
$\quad\quad = 1.20 > 1.0 \quad\quad$ Eliminate DN.

<u>B to A:</u>
$\Delta C = (90 - 50)(A/P,10\%,5) + (4 - 3)$
$\quad = 40(0.26380) + 1$
$\quad = \$11.552$

$$\triangle B = (29 - 20) - (1.5 - 0.5)$$
$$= 8$$
$$B/C = \frac{8}{11.552}$$
$$= 0.69 < 1.0 \qquad \text{Eliminate B.}$$

C to A
$$\triangle C = (200 - 50)(A/P, 10\%, 5) + (6 - 3)$$
$$= 150(0.26380) + 3$$
$$= 42.57$$
$$\triangle B = (61 - 20) - (2.1 - 0.5)$$
$$= 39.4$$
$$B/C = \frac{39.4}{42.57}$$
$$= 0.93 < 1.0 \qquad \text{Eliminate C.}$$

Select site A.

9.28 (a) An incremental B/C analysis is necessary between Y and Z, if these are mutually exclusive alternatives.

(b) Independent projects. Accept Y and Z, since B/C > 1.0.

9.29 J to DN
B/C = 1.10 > 1.0   Eliminate DN.

K to J
B/C = 0.40 < 1.0   Eliminate K.

L to J
B/C = 1.42 > 1.0   Eliminate J.

M to L
B/C = 0.08 < 1.0   Eliminate M.

Select alternative L.

Note: K and M can be eliminated initially because they have B/C < 1.0.

9.30 (a) Projects are listed by increasing PW of cost values. First find benefits for each alternative and then find incremental B/C ratios:

Benefits for P
$$1.1 = B_P / 10$$
$$B_P = 11$$

Chapter 9 - 11 -

Benefits for Q
$2.4 = B_Q/40$
$B_Q = 96$
Benefits for R
$1.4 = B_R/50$
$B_R = 70$
Benefits for S
$1.5 = B_S/80$
$B_S = 120$
Incremental B/C for Q vs P:
$$B/C = \frac{96 - 11}{40 - 10} = 2.83$$
Incremental B/C for R vs P:
$$B/C = \frac{70 - 11}{50 - 10} = 1.48$$
Incremental B/C for S vs P:
$$B/C = \frac{120 - 11}{80 - 10} = 1.56$$
Incremental B/C for R vs Q:
$$B/C = \frac{70 - 96}{50 - 40} = -2.60$$
Disregard due to less B for more C.
Incremental B/C for S vs Q:
$$B/C = \frac{120 - 96}{80 - 40} = 0.60$$
Incremental B/C for S vs R:
$$B/C = \frac{120 - 70}{80 - 50} = 1.67$$

(b) Compare P to DN; eliminate DN.
Compare Q to P; eliminate P.
Compare R to Q; disregarded.
Compare S to Q; eliminate S.
Select Q.

FE Review Solutions

9.31   Answer is (d)

9.32   Answer is (b)

9.33   Answer is (a)

9.34   Answer is (b)

9.35   Answer is (c)

Extended Exercise solution

1. The spreadsheet shows the incremental B/C analysis. The truck should be purchased. The annual worth values for each alternative are determined using the equations:

$$AW_{pay\text{-}per\text{-}use} = 150{,}000(A/P,6\%,5) + 10(3000) + 3(8000) = \$89{,}609 \text{ (cell D15)}$$

$$AW_{own} = 850{,}000(A/P,6\%,15) + 500{,}000(A/P,6\%,50) + 15(2000) + 5(7000) = \$184{,}240 \text{ (cell F15)}$$

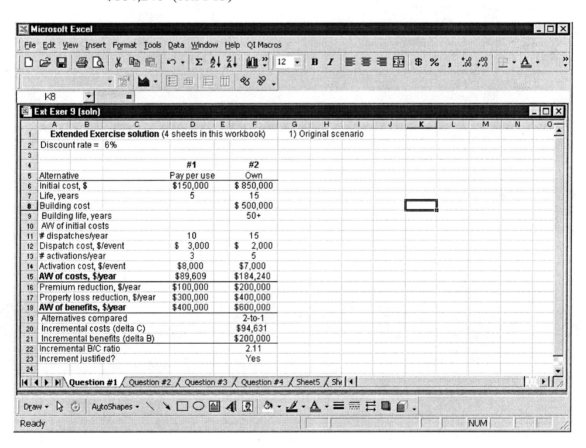

2. The annual fee paid for 5 years now would have to be negative (cell D5) in that Brewster would have to pay Medford a 'retainer fee', so to speak, to possibly use the ladder truck. This is an economically unreasonable approach.

Excel SOLVER is used to find the breakeven value of the initial cost when B/C = 1.0 (cell F21).

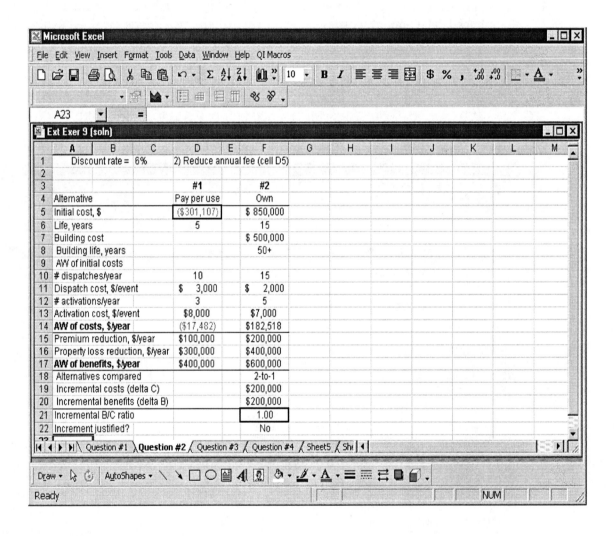

Chapter 9

3. The building cost of over $2.2 million could be supported by the Brewster proposal (in cell F7), again found by using SOLVER. This is also not an economically reasonable alternative.

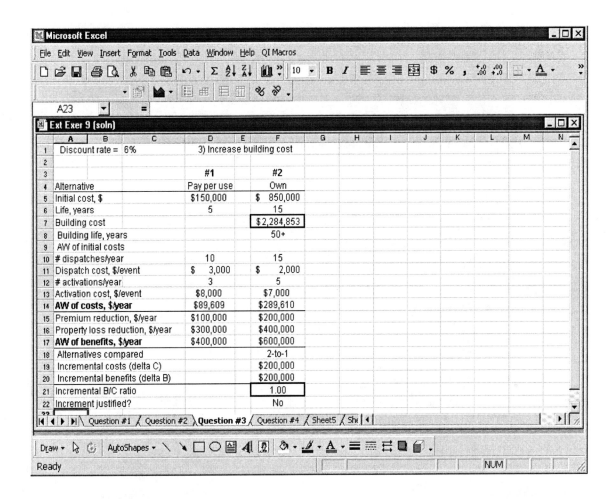

4. The estimated sum of premium and property loss would need to be $523,714 or less (cell F17, SOLVER). This is not much of a reduction from the current estimate of $600,000.

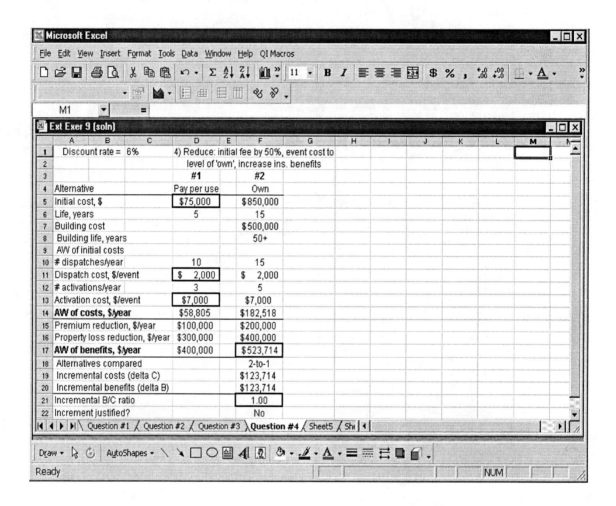

Case Study Solution

1. Installation cost = (3,500)[87.8/(0.067)(2)]
   = (3,500)(655)
   = $2,292,500

   Annual power cost = (655 poles)(2)(0.4)(12)(365)(0.08)
   = $183,610

   Total annual cost = 2,292,500(A/P,6%,5) + 183,610
   = $727,850

   If the accident reduction rate is assumed to be the same as that for closer spacing of lights,

   B/C = 1,111,500/727,850
   = 1.53

2. Night/day deaths, unlighted = 5/3 = 1.6

   Night/day deaths, lighted = 7/4 = 1.8

3. Installation cost = 2,500(87.8/0.067)
   = $3,276,000

   Total annual cost = 3,276,000(A/P, 6%, 5) + 367,219
   = $1,144,941

   B/C = 1,111,500/1,144,941
   = 0.97

4. Ratio of night/day accidents, lighted = $\frac{839}{2069}$ = 0.406

   If the same ratio is applied to unlighted sections, number of accidents prevented where property damage was involved would be calculated as follows:

   0.406 = $\frac{\text{no. of accidents}}{379}$

   no. accidents = 154

   no. prevented = 199 – 154 = 45

Chapter 9

5. For lights to be justified, benefits would have to be at least $1,456,030 (instead of $1,111,500). Therefore, the difference in the number of accidents would have to be:

   1,456,030 = (difference)(4500)
   Difference = 324

   No. of accidents would have to be = 1086 − 324 = 762

   Night/day ratio = $\dfrac{762}{2069}$ = 0.368

Chapter 10 – Making Choices: the Method, MARR, and Multiple Attributes
Solutions to end of chapter exercises

Problems

10.1 The circumstances are when the lives for all alternatives are: (1) finite and equal, or (2) considered infinite. It is also correct when (3) the evaluation will take place over a specified study period.

10.2 Incremental cash flow analysis is mandatory for the ROR method and B/C method. (It is noteworthy that if unequal-life cash flows are evaluated by ROR using an AW-based relation that reflects the differences in cash flows between two alternatives, the breakeven i^* will be the same as the $\triangle i^*$. (See Table 10.2 and Section 10.1 for comments.)

10.3 Numerically largest means the alternative with the largest PW, AW or FW identifies the selected alternative. For both revenue and service alternatives, the largest number is chosen. For example, $-5000 is selected over $-10,000, and $+100 is selected over $-50.

10.4 (a) Hand solution: After consulting Table 10.1, choose the AW or PW method at 8% for equal lives of 8 years.

Computer solution: either the PMT function or the PV function can give single-cell solutions for each alternative.

In either case, select the alternative with the numerically largest value of AW or PW.

(b) Hand solution: Find the PW for each cash flow series.

PW_8 = -10,000 + 2000(P/F,18%,8) + (6500 – 4000) (P/A,18%,8)
    = -10,000 + 2000(0.2660) + 2500(4.0776)
    = $726

PW_{10} = -14,000 + 2500(P/F,18%,8) + (10,000 – 5500) (P/A,18%,8)
     = $5014

PW_{15} = -18,000 + 3000(P/F,18%,8) + (14,000 – 7000) (P/A,18%,8)
     = $11,341

PW_{20} = -24,000 + 3500(P/F,18%,8) + (20,500 – 11,000) (P/A,18%,8)
     = $15,668

$$PW_{25} = -33{,}000 + 6600(P/F,18\%,8) + (26{,}500 - 16{,}000)(P/A,18\%,8)$$
$$= \$11{,}570$$

Select the 20 cubic meter size.

Computer solution: Use the PV function to find the PW in a separate spreadsheet cell for each alternative. Select the 20 cubic meter alternative.

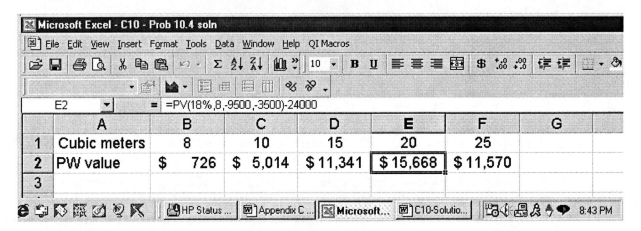

10.5 (a) Hand solution: Choose the AW or PW method at 0.5% for equal lives over 60 months.

Computer solution: Either the PMT function or the PV function can give single-cell solutions for each alternative.

(b) Hand solution: Find the AW for each cash flow series on a per household per month basis.

$$AW_1 = 1.25 - 60(A/P,0.5\%,60)$$
$$= 1.25 - 60(0.01933)$$
$$= 1.25 - 1.16$$
$$= \$0.09$$

$$AW_2 = 8.00 - 500(A/P,0.5\%,60)$$
$$= 8.00 - 9.67$$
$$= \$-1.67$$

Select program 1.

Computer solution: Develop the AW value using the PMT function in a separate cell for each program. Select program 1.

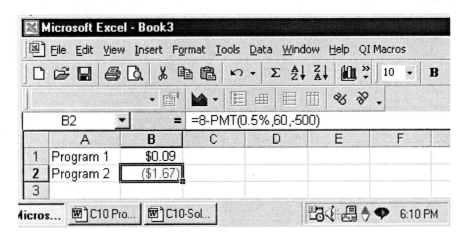

10.6 (a) The expected return is 15 - 10 = 5% per year.
 (b) Retain MARR = 15% and then estimate the project i*. Take the risk-related return expectation into account before deciding on the project. If 15% < i* < 20%, John must decide if the risk is worth less than 5% over MARR = 15%.

10.7 (a) Bonds are debt financing
 (b) Stocks are always equity
 (c) Equity
 (d) Mortgage loans are debt

10.8 The project that is rejected, say B, and has the next highest ROR measure, i^*_B, in effect sets the MARR, because it's rate of return is a lost opportunity rate of return. Were a second alternative selected, B would be it and the effective MARR would have been i^*_B.

10.9 (a) Select C. It is the alternative investing the maximum available with $\triangle i^* > 9\%$.

 (b) The first rejected alternative within the capital funds available is B; the opportunity cost is 12.5% per year.

10.10 Set the MARR at the cost of capital. Determine the rate of return for the cash flow estimates and select the best alternative. Examine the difference between the return and MARR to separately determine if it is large enough to cover the other factors for this selected alternative. (This is different than increasing the MARR before the evaluation to accommodate the factors.)

10.11 (a) MARR may tend to be set lower, based on the success of the last purchase.
(b) Set the MARR and then treat the risk associated with the purchase separately from the MARR.

10.12 (a) Calculate the two WACC values.

$$WACC_1 = 0.4(9\%) + 0.6(10\%) = 9.6\%$$
$$WACC_2 = 0.25(9\%) + 0.75(10.5\%) = 10.125\%$$

Use approach 1, with a D-E mix of 40%-60%

(b) Let x_1 and x_2 be the maximum costs of debt capital.

Alternative 1: $10\% = WACC_1 = 0.4(9\%) + 0.6(x_1)$
$x_1 = [10\% - 0.4(9\%)]/0.6$
$= 10.67\%$

Debt capital cost could increase from 10% here.

Alternative 2: $10\% = WACC_2 = 0.25(9\%) + 0.75(x_2)$
$x_2 = [10\% - 0.25(9\%)]/0.75$
$= 10.33\%$

Debt capital cost would have to go down from 10.5% here.

10.13 (a) Debt: $135,000 total from the mortgage loan for $110,000 at 8%, and grandfather loan for $25,000 at 0%.

Debt percentage = 135,000/150,000 = 90%

Equity: investment savings of $15,000 at 11%.

Equity percentage = 15,000/150,000 = 10%

D-E mix = 90%-10%

In $1,000 units,

$$WACC = \frac{\$15(11\%) + 25(0\%) + 110(8\%)}{\$150}$$

$$= (165 + 0 + 880) / 150$$
$$= 6.97\%$$

(b) Same D-E mix of 90%-10% with a mortgage loan of $135,000 at 8%. WACC changes. In $1,000 units:

$$WACC = \frac{\$15(11\%) + 135(8\%)}{\$150}$$

$$= (165 + 1080) / 150$$
$$= 8.3\%$$

10.14 WACC = cost of debt capital + cost of equity capital
= (0.7)[0.6(11%) + 0.4(9%)] + (0.3)[(0.75)(5%) + (0.25)(14%)]
= 0.7[10.2%] + 0.3[7.25%]
= 9.32%

10.15 (a) Compute and plot WACC for each D-E mix.

D-E Mix	WACC
0-100	14.50%
30-70	11.44
35-65	10.25
65-35	9.84
80-20	12.72
100-0	11.50

Take the equity and debt capital data from the problem statement.

(b) D-E mix of 65-35 has the lowest WACC values over the period of time.

10.16 Solve for the debt cost of capital, x.

$$WACC = 10.7\% = 0.8(6\%) + (1-0.8)(x)$$
$$x = (10.7 - 4.8)/0.2$$
$$= 29.5\%$$

The rate of 29.5% for debt capital (loans, bonds, etc.) is very high.

10.17 The before-tax WACC is

$$WACC = 0.5(10\%) + 0.5(14\%) = 12\% \text{ per year}$$

After-tax WACC = 12%(1-0.38)
= 7.44% per year

10.18 (a) Face value = $\frac{\$1,200,000}{0.98}$ = $1,224,490

(b) Bond interest = $\frac{0.074(1,224,490)}{2}$ = $45,306 every 6 months

Interest net cash flow = $45,306(1 - 0.4) = $27,184

The rate of return equation per 6-months is:

$0 = 1,200,000 - 27,184(P/A,i^*,30) - 1,224,490(P/F,i^*,30)$

$i^* = 2.31\%$ per 6 months   (RATE or IRR function)

Nominal $i^* = 4.62\%$ per year

10.19 Equity cost of capital is stated as 6%. Debt cost of capital benefits from tax savings.

Before-tax bond annual interest = $4 million (0.08) = $320,000
Annual bond interest NCF = $320,000(1 – 0.4) = $192,000
Effective quarterly dividend = $48,000

Find quarterly i^* for the PW relation.

$0 = 4,000,000 - 48,000(P/A,i^*,40) - 4,000,000(P/F,i^*,40)$

$i^* = 1.2\%$ per quarter
= 4.8% per year (nominal)

Debt financing at 4.8% per year is cheaper than equity funds at 6% per year.

(Note: The correct answer is also obtained if the before-tax debt cost of 8% is used to estimate the after-tax debt cost of 8%(1 - 0.4) = 4.8%.)

10.20 (a) <u>Bank loan:</u>
Annual loan payment = 800,000(A/P,8%,8)
= 800,000(0.17401)
= $139,208
Principal payment = 800,000/8 = $100,000
Annual interest = 139,208 – 100,000 = $39,208

Tax saving = 39,208(0.40) = $15,683
Effective interest payment = 39,208 – 15,683 = $23,525
Effective annual payment = 23,525 + 100,000 = $123,525

The AW-based i* relation is:

$0 = 800,000(A/P,i^*,8) - 123,525$

$(A/P,i^*,8) = \dfrac{123,525}{800,000} = 0.15441$

$i^* = 4.95\%$

Bond issue:
Annual bond interest = 800,000(0.06) = $48,000
Tax saving = 48,000(0.40) = $19,200
Effective bond interest = 48,000 – 19,200 = $28,800

The AW-based i* relation is:

$0 = 800,000(A/P,i^*,10) - 28,800 - 800,000(A/F,i^*,10)$

$i^* = 3.6\%$ (RATE or IRR function)

Bond financing is cheaper.

(b) Bonds cost 6% per year, which is less than the 8% loan. The answer is the same before-taxes.

10.21 Face value of bond issue = (10,000,000)/ 0.975 = $10,256,410

Annual bond interest = 0.0975(10,256,410) = $1,000,000

Interest net cash flow = $1,000,000(1 - 0.32) = $680,000

The PW-based rate of return equation is:

$0 = 10,000,000 - 680,000(P/A,i^*,30) - 10,256,410(P/F,i^*,30)$

$i^* = 6.83\%$ per year   (RATE or IRR function)

Bonds are cheaper than the bank loan at 7.5% with no tax advantage.

10.22 Dividend method:

$$R_e = DV_1/P + g$$
$$= 1.25/25.50 + 0.03$$
$$= 7.9\%$$

CAPM: (The return values are in percents.)

$$R_e = R_f + \beta(R_m - R_f)$$
$$= 7.5 + 0.9(9.5 - 7.5)$$
$$= 9.3\%$$

CAPM estimate is 1.4% higher.

10.23 Debt capital cost: 9.5% for $6 million (60% of total capital)

Equity -- common stock: 100,000(32) = $3.2 million or 32% of total capital

$$R_e = 1.10/32 + 0.02$$
$$= 5.44\%$$

Equity -- retained earnings: cost is 5.44% for this 8% of total capital.

$$WACC = 0.6(9.5\%) + 0.32(5.44\%) + 0.08(5.44\%)$$
$$= 7.88\%$$

10.24 Determine the effective annual interest rate i_a for each plan. All the dollar values can be neglected.

Plan A:
$$i_a \text{ for debt} = (1 + 0.0075)^{12} - 1 = 9.381\%$$
$$i_a \text{ for equity} = (1 + 0.0175)^{4} - 1 = 7.186\%$$

$$WACC_A = 0.5(9.381\%) + 0.5(7.186\%) = 8.28\%$$

Plan B:
$$i_a \text{ for 100\% equity} = WACC_B = (1 + 0.0175)^{4} - 1 = 7.186\%$$

Plan C:
$$i_a \text{ for 100\% debt} = WACC_C = (1 + 0.0075)^{12} - 1 = 9.381\%$$

Plan B: 100% equity has the lowest WACC.

10.25 (a) Equity capital: 50% of capital at 15% per year.

Debt capital: 15% in bonds and 35% in loans.

Cost of loans: 10.5% per year

Cost of bonds: 6% from the problem statement, or determine i^*.

Bond annual interest per bond = $10,000(0.06) = $600

$0 = 10,000 - 600(P/A,i^*,15) - 10,000(P/F,i^*,15)$

$i^* = 6.0\%$ (RATE or IRR function)

WACC = 0.5(15%) + 0.15(6%) + 0.35(10.5%)
 = 12.075%

(b) Use T_e = 35% to calculate after-tax WACC using Eq. [10.3].

After-tax WACC = (before-tax WACC)(1- T_e)
 = 12.075% (1-0.35)
 = 7.85%

10.26 Use Eq. [10.7] with R_f = 5.19% and the 1-year return amount as R_m.

Equity capital: Vanguard Aggressive Growth Stock

R_e = 5.19 + 1.06(25.24 - 5.19)
 = 26.44%

Equity capital: Standard and Poor's 500 Stocks

R_e = 5.19 + 1.00(17.94 - 5.19)
 = 17.94%

Equity capital: Vanguard Equity Income Stock

R_e = 5.19 + 0.75(13.18 - 5.19)
 = 11.18%

Debt capital (the safe investment index): Salomon 3-month U. S. Treasury Bill

$$R_e = 5.19 + 1.00(5.19 - 5.19)$$
$$= 5.19\%$$

Debt capital (comparison index): Vanguard Short term US Treasury Bond

$$R_e = 5.19 + 0.51(5.85 - 5.19)$$
$$= 5.53\%$$

10.27 For the D-E mix of 70%-30%, WACC = 0.7(7.0%) + 0.3(10.34%) = 8.0%

$$MARR = WACC = 8\%$$

(a) Independent projects: These are revenue projects. Fastest solution is to find PW at 8% for each project. Select all those with PW > 0.

$$PW_1 = -25,000 + 6,000 \ (P/A,8\%,4) + 4,000 \ (P/F,8\%,4)$$

$$PW_2 = -30,000 + 9,000 \ (P/A,8\%,4) - 1,000 \ (P/F,8\%,4)$$

$$PW_3 = -50,000 + 15,000 \ (P/A,8\%,4) + 20,000 \ (P/F,8\%,4)$$

Spreadsheet solution below shows PW at 8% and overall i*.

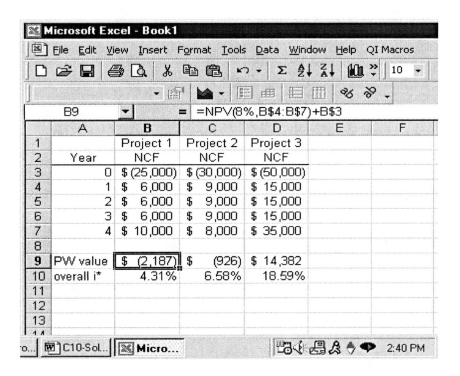

Only project 3 has PW > 0. Select it.

(b) Mutually exclusive: Since only $PW_3 > 0$, select it.

10.28 One approach is to utilize a >cost only= analysis and incrementally compare alternatives against each other with out the possibility of selecting the do-nothing alternative.

10.29 A large D-E mix over time is not healthy financially because this indicates that the company owns too small of a percentage of itself (equity ownership) and is a risky investment for lenders. Under these circumstances, in tight money situations, cash and debt capital will be hard to obtain and very expensive in terms of interest rates.

10.30 <u>100% equity financing</u>

MARR = 8.5% is known. Determine PW at the MARR.

$$
\begin{aligned}
PW &= -250{,}000 + 30{,}000(P/A,8.5\%,15) \\
&= -250{,}000 + 30{,}000(8.3042) \\
&= -250{,}000 + 249{,}126 \\
&= \$\text{-}874
\end{aligned}
$$

Since PW < 0, 100% equity does not meet the MARR requirement.

60%-40% D-E financing

Loan principal = 250,000(0.60) = $150,000
Loan payment = 150,000(A/P,5%,15)
= 150,000(0.09634)
= $14,451 per year

Cost of debt capital is 5% for the loan.

WACC = 0.4(8.5%) + 0.6(5%) = 6.4%
MARR = 6.4%

Annual NCF = project NCF - loan payment
= $30,000 - 14,451 = $15,549

Amount of equity invested = 250,000 - 150,000 = $100,000

Calculate PW at the MARR on the basis of the committed equity capital.

PW = -100,000 + 15,549(P/A,6.4%,15)
= -100,000 + 15,549(9.4634)
= $47,146

Conclusion: PW > 0; select the 60%-40% debt-equity financing option.

10.31 Determine i^* for each plan.

Plan 1: 90% equity means $450,000 funds are invested. Use a PW-based relation.

0 = -450,000 + 100,000 (P/A,i^*,5)
i_1^* = 3.62%   (RATE function)

Plan 2: 60% equity means $300,000 invested.

0 = -300,000 + 100,000 (P/A,i^*,5)
i_2^* = 19.86%   (RATE function)

Plan 3: 20% equity means $100,000 invested.
0 = -100,000 + 100,000(P/A,i^*,5)
i_3^* = 96.59%   (RATE function)

Calculate the MARR values.

(a) MARR = 10% all plans

(b) $MARR_1 = WACC_1 = 0.9(10\%) + 0.1(12\%) = 10.2\%$
$MARR_2 = WACC_2 = 0.6(10\%) + 0.4(12\%) = 10.8\%$
$MARR_3 = WACC_3 = 0.2(10\%) + 0.8(12\%) = 11.6\%$

(c) $MARR_1 = (10.2 + 10.0)/2 = 10.1\%$
$MARR_2 = (10.8 + 10.0)/2 = 10.4\%$
$MARR_3 = (11.6 + 10.0)/2 = 10.8\%$

Make the decisions using i* values for each plan.

Plan	i*	Part (a) MARR	?+	Part (b) MARR	?+	Part (c) MARR	?+
1	3.62%	10%	N	10.2%	N	10.1%	N
2	19.86	10	Y	10.8	Y	10.4	Y
3	96.59	10	Y	11.6	Y	10.8	Y

(+Table legend: >?= is for the question "Is the plan justified in that i* > MARR?")

Same decision for all 3 options; plans 2 and 3 are acceptable.

10.32 (a) Find cost of equity capital from CAPM.

$R_e = 8\% + 1.05(5\%) = 13.25\%$

MARR = 13.25%

Find i* on 50% equity investment.

$0 = -5,000,000 + 2,000,000(P/A,i^*,6)$
i* = 32.66%   (RATE function)

The investment is economically acceptable since i* > MARR.

(b) Determine WACC and set MARR = WACC. For 50% debt financing at 8%,

WACC = MARR = 0.5(8%) + 0.5(13.25%) = 10.63%

Chapter 10

The investment is acceptable, since 32.66% > 10.63%.

10.33 All points will increase, except the 0% debt value. The new WACC curve is relatively higher at both the 0% debt and 100% debt points and the minimum WACC point will move to the right.

Conclusion: The minimum WACC will increase with a higher D-E mix, since debt and equity cost curves rise relative to those for lower D-E mixes.

10.34 Ratings by attribute with 10 for #1.

Attribute	Importance	Logic
1	10	Most important (10)
2	2.5	0.5(5) = 2.5
3	5	1/2(10) = 5
4	5	2(2.5) = 5
5	5	2(2.5) = 5
	27.5	

W_i = Score/27.5

Attribute	W_i
1	0.364
2	0.090
3	0.182
4	0.182
5	0.182
	1.000

10.35 Lease cost (as an alternative to purchase)
Insurance cost
Resale value
Safety features
Pick-up (acceleration)
Steering response
Quality of ride
Aerodynamic design
Options package
Cargo volume
Warranty
What friends own

10.36 Calculate W_i = Importance score/sum and use Eq. [10.11] for R_j.

Vice president

Attribute, i	Importance score	W_i	V_{ij} values 1	2	3
1	20	0.10	5	7	10
2	80	0.40	40	24	12
3	100	0.50	50	20	25
	Sum = 200		95	51	47 = R_j values

Select Alternative 1 since R_1 is largest.

Assistant vice president

Attribute, i	Importance score	W_i	V_{ij} values 1	2	3
1	100	0.50	25	35	50
2	80	0.40	40	24	12
3	20	0.10	10	4	5
	Sum = 200		75	63	67 = R_j values

With $R_1 = 75$, select alternative 1

Results are the same, even though the VP and asst.VP rated opposite on factors 1 and 3. High score on attribute 1 by asst.VP is balanced by the VP's score on attributes 2 and 3.

10.37 (a) Both sets of ratings give the same conclusion, alternative 1, but the consistency between raters should be improved somewhat. This result simply shows that the weighted evaluation method is relatively insensitive to attribute weights when an alternative (1 here) is favored by high (or disfavored by low) weights.

(b) ### Vice president

Take W_i from problem 10.36. Calculate R_j using Eq. [10.11].

Attribute	W_i	V_{ij} 1	2	3
1	0.10	3	4	10
2	0.40	28	40	28
3	0.50	50	40	45
		81	84	83

Conclusion: Select alternative 2.

Assistant vice president

Attribute	W_i	V_{ij} for alternatives 1	2	3
1	0.50	15	20	50
2	0.40	28	40	28
3	0.10	10	8	9
		53	68	87

Conclusion: Select 3.

(c) There is now a big difference for the asst. VP's alternative 3 and the VP has a very small difference between alternatives. The VP could very easily select alternative 3, since the R_j values are so close.

Reverse rating of VP and asst.VP makes only a small difference in choice, but it shows real difference in perspective. Rating difference in alternatives by attribute can make a significant difference in the alternative selected, based on results in this problem.

10.38 Sum the ratings in Table 10.5 over all six attributes.

	V_{ij} 1	2	3
Total	470	515	345

Select alternative 2; the same choice is made.

10.39 (a) Select A since PW is larger.

(b) Use Eq. [10.11] and manager scores for attributes.

$$W_i = \frac{\text{Importance score}}{\text{Sum}}$$

Attribute, i	Importance (By mgr.)	W_i	R_j A	B
1	100	0.57	0.57	0.51
2	35	0.20	0.07	0.20
3	20	0.11	0.11	0.10
4	20	0.11	0.03	0.11
	175		0.78	0.92

Select B.

(c) Use Eq. [10.11] and trainer scores for attributes.

Attribute i	Importance (By trainer)	W_i	R_j A	B
1	80	0.40	0.40	0.36
2	10	0.05	0.02	0.05
3	100	0.50	0.50	0.45
4	10	0.05	0.01	0.05
	200		0.93	0.91

Select A, by a small margin

Note: 2 methods indicate A and 1 indicates B.

Extended Exercise Solution

1. Use scores as recorded to determine weights by Equation [10.10]. Note that the scores are not rank ordered, so a 1 indicates the most important attribute. Therefore the *lowest weight is the most important attribute*. The sum or average can be used to find the weights.

Attribute	Committee member 1	2	3	4	5	Sum	Avg.	W_j
A. Closeness to the citizenry	4	5	3	4	5	21	4.2	0.280
B. Annual cost	3	4	1	2	4	14	2.8	0.186
C. Response time	2	2	5	1	1	11	2.2	0.147
D. Coverage area	1	1	2	3	2	9	1.8	0.120
E. Safety of officers	5	3	4	5	3	20	4.0	0.267
Totals						75	15.0	1.000

W_j = sum/75 = average/15

For example,

$W_1 = 21/75 = 0.280$ or $W_1 = 4.2/15 = 0.280$

$W_2 = 14/75 = .186$ or $W_2 = 2.8/15 = 0.186$

2. Attributes B, C, and D are retained. (The 'people factor' attributes have been removed.) Renumber the remaining attributes in the same order with scores of 1, 2, and 3.

	Committee member							
Attribute	1	2	3	4	5	Sum	Avg.	W_j
B. Annual cost	3	3	1	2	3	12	2.4	0.4
C. Response time	2	2	3	1	1	9	1.8	0.3
D. Coverage area	1	1	2	3	2	9	1.8	0.3
Totals						30	6.0	1.0

Now, W_j = sum/30 = average/6

3. Because the most important attribute lowest score of 1, select the *two smallest R_j values* in question 1. Therefore, the chief should choose the <u>horse and foot options</u> for the pilot study.

Case Study Solution

1. Set MARR = WACC

   WACC = (% equity)(cost of equity) + (% debt)(cost of debt)

   Equity   Use Eq. [10.6]

   $$R_e = \frac{0.50}{15} + 0.05 = 8.33\%$$

   Debt   Interest is tax deductible; use Eqs. [10.4] and [10.5].

   Tax savings = Interest(tax rate)
   = [Loan payment – principal portion](tax rate)

   Loan payment = 750,000(A/P,8%,10) = $111,773 per year

   Interest = 111,773 – 75,000 = $36,773

   Tax savings = (36,773)(.35) = $12,870

   Cost of debt capital is i* from a PW relation:

   0 = loan amount – (annual payment after taxes)(P/A,i*,10)
   = 750,000 – (111,773 – 12,870)(P/A,i*,10)

   (P/A,i*,10) = 750,000 / 98,903 = 7.5832

   i* = 5.37%   (RATE function)

   Plan A(50-50): MARR = WACC$_A$ = 0.5(5.37) + 0.5(8.3) = 6.85%

   Plan B(0-100%): MARR = WACC$_B$ = 8.33%

2.   A: <u>50–50 D–E financing</u>
   Use relations in case study statement and the results from Question #1.

   TI = 300,000 – 36,773 = $263,227
   Taxes = 263,227(0.35) = $92,130

   After-tax NCF = 300,000 – 75,000 – 36,773 – 92,130
                   = $96,097

Find plan i_A^* from AW relation for $750,000 of equity capital

$0 = $ (committed equity capital)$(A/P, i_A^*, n) + S(A/F, i_A^*, n) +$ after tax NCF

$0 = -750,000(A/P, i_A^*, 10) + 200,000(A/F, i_A^*, 10) + 96,097$

$i_A^* = 7.67\%$ (RATE function)

Since $7.67\% > \text{WACC}_A = 6.85\%$, plan A is acceptable.

B: <u>0–100 D–E financing</u>
Use relations is the case study statement

After tax NCF $= 300,000(1–0.35) = \$195,000$
All $1.5 million is committed. Find i_B^*

$0 = -1,500,000(A/P, i_B^*, 10) + 200,000(A/F, i_B^*, 10) + 195,000$

$i_B^* = 6.61\%$ (RATE function)

Now $6.61\% < \text{WACC}_B = 8.33\%$, plan B is rejected.

Recommendation: Select plan A with 50-50 financing.

3. Spreadsheet shows the hard way (develops debt-related cash flows for each year, then obtains WACC) and the easy way (uses costs of capital from #1) to plot WACC.

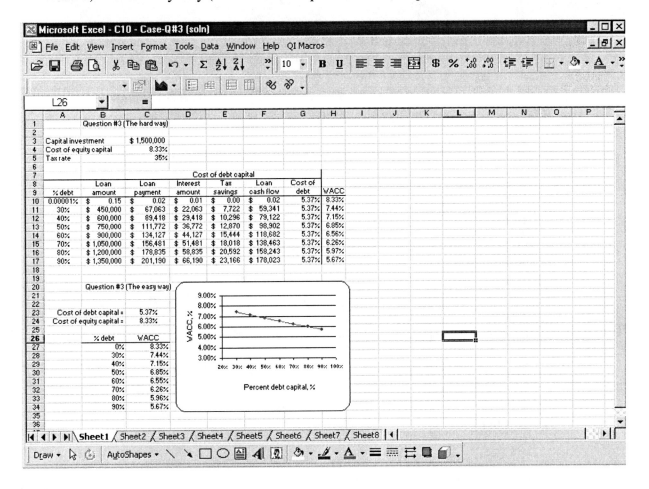

Chapter 11 – Replacement and Retention Decisions
Solutions to end of chapter exercises

Problems

11.1 A consultant's viewpoint is that of an unbiased observer who would not be influenced by whether or not an asset is presently owned.

11.2 The value of P that should be used for a currently-owned asset is its market value. If the asset must be updated or augmented, this cost is added to the first cost.

11.3 The defender/challenger refers to the currently-owned asset (the defender) and the one that might be its replacement (challenger).

11.4 Reasons why a replacement study might be needed are:
(1) Reduced performance (2) altered requirements (3) obsolescence

11.5 Assumptions commonly made in a replacement study are:
(1) Services are needed for forever.
(2) Challenger is best available now and it will be repeated in succeeding life cycles.
(3) Cost of challenger will be same in all life cycles.

11.6 P = market value = $40,000
AOC = $15,000 per year
n = 4 years
S = $25,000

11.7 (a) P = 40,000 – 3000(2) = $34,000
S = 40,000 – 3000(3) = $31,000
AOC = 32,000 + 1000k

(b) P = 40,000 – 3000(3) = $31,000
S = 40,000 – 3000(4) = $28,000
AOC = 33,000 + 1000k

11.8 P = MV = 85,000 – 10,000(1) = $75,000
AOC = $36,500 + 1,500k (k = 1 to 5)
n = 5 years
S = 85,000 – 10,000(6) = $25,000

11.9 Set up AW equations for 1 through 5 years and solve by hand or spreadsheet (PMT function).

For n=1:   AW = -70,000(A/P,20%,1) – 20,000 + 10,000(A/F,20%,1)
              = -70,000(1.2000) – 20,000 + 10,000(1.0000)
              = $-94,000

For n=2: AW = -70,000(A/P,20%,2) – 20,000 + 10,000(A/F,20%,2)
          = $-61,273

For n=3: AW = $-50,484

For n=4: AW = $-45,177

For n=5: AW = $-42,063

Economic service life is 5 years with AW = $-42,063.

11.10 (a) Set up AW equations for each year 1 through 10 and solve by hand or PMT function on a spreadsheet.

For n=1: AW = -50,000(A/P,15%,1) – 25,000 + 40,000(A/F,15%,1)
          = $-42,500

For n=2: AW = -50,000(A/P,15%,2) – 25,000 + 30,000(A/F,15%,2)
          = $-41,802

For n=3: AW = -50,000(A/P,15%,3) – 25,000 + 20,000(A/F,15%,3)
          = $-41,139

For n=4: AW = 50,000(A/P,15%,4) – 25,000 + 18,000(A/F,15%,3)
          = $-38,908

For n=5: AW = $-37,543

For n=6: AW = $-36,613

For n=7: AW = $-35,934

For n=8: AW = $-35,414   (using PMT function)

For n=9: AW = $-35,002

For n=10: AW = $-34,667

Economic service life is 10 years with AW = $-34,667.

(b) Use the PMT function with S = $40,000 and S = 0 for each year. The ESL remains at 10 years for all values and combinations.

11.11 Set up AW equations for years 1-6 and solve by hand or PMT function.

For n=1: $AW = -60,000(A/P,18\%,1) - 75,000 + 60,000(1 - 0.15)^1 (A/F,18\%,1)$
$= \$-94,800$

For n=2: $AW = -60,000(A/P,18\%,2) - 75,000 + 60,000(1 - 0.15)^2 (A/F,18\%,2)$
$= \$-93,438$

For n=3: $AW = \$-92,281$

For n=4: $AW = \$-91,299$

For n=5: $AW = \$-90,466$

For n=6: $AW = \$-89,758$

Economic service life is 6 years with $AW = \$-89,758$.

11.12 Set up AW equations for years 1-5 and solve by hand or Excel.

For n=1: $AW = -90,000(A/P,20\%,1) - 50,000 + 60,000(A/F,20\%,1)$
$= \$-98,000$

For n=2: $AW = \$-90,000(A/P,20\%,2) - [50,000 + 5000(A/G,20\%,2)] + 50,000(A/F,20\%,2)$
$= \$-88,454$

For n=3: $AW = \$-86,132$

For n=4: $AW = \$-85,548$

For n=5: $AW = \$-85,609$

Economic service life is 4 years with $AW = \$-85,548$.

The spreadsheet commands for the total AW of costs are shown below:

Chapter 11

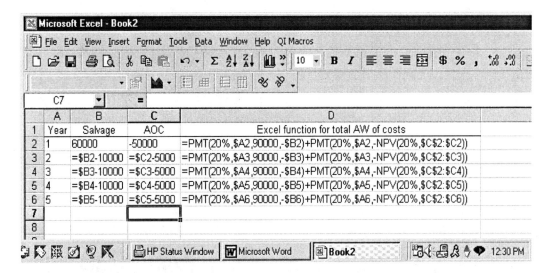

11.13 Set up AW equations for n=1 through 8 and solve by hand or spreadsheet:

For n=1: AW = -150,000(A/P,15%,1) – 72,000 + 100,000(A/F,15%,1)
 = $-144,500

For n=2: AW = -150,000(A/P,15%,2) – [72,000 + 2000(A/G,15%,2)]
 + 80,000(A/F,15%,2)
 = $-127,988

For n=3: AW = $-122,232

For n=4: AW = $-119,182

For n=5: AW = $-117,227

For n=6: AW = $-115,830

For n=7: AW = $-112,954

For n=8: AW = $-110,990

Economic service life is 8 years with AW = $-110,990.

(b) Spreadsheet below shows all the same results as the manual solution above.
 ESL is 8 years with AW = $-110,990.

Chapter 11 - 4 -

	A	B	C	D	E	F	G	H
1								
2		Market	Loss in MV	Lost interest		MC for	AW of	
3	Year	value	for year	MV for year	AOC	year	marginal cost	
4	0	$150,000						
5	1	$100,000	$ 50,000	$ 22,500	$72,000	$144,500	$ (144,500)	
6	2	$ 80,000	$ 20,000	$ 15,000	$74,000	$109,000	$ (127,988)	
7	3	$ 60,000	$ 20,000	$ 12,000	$76,000	$108,000	$ (122,232)	
8	4	$ 40,000	$ 20,000	$ 9,000	$78,000	$107,000	$ (119,182)	
9	5	$ 20,000	$ 20,000	$ 6,000	$80,000	$106,000	$ (117,227)	
10	6	$ -	$ 20,000	$ 3,000	$82,000	$105,000	$ (115,830)	
11	7	$ -	$ -	$ -	$84,000	$ 84,000	$ (112,954)	
12	8	$ -	$ -	$ -	$86,000	$ 86,000	$ (110,990)	ESL
13								
14								

11.14 Set up AW equations for n = 1 through 7 and solve by hand or spreadsheet:

For n=1: AW = -100,000(A/P,14%,1) – 28,000 + 75,000(A/F,14%,1)
                = $-67,000

For n=2: AW = -100,000(A/P,14%,2) - [28,000(P/F,14%,1) + 31,000
                (P/F,14%,2)] (A/P,14%,2) + 60,000(A/F,14%,2)
                = $-62,093

For n=3: AW = $-59,275

For n=4: AW = $-57,594

For n=5: AW = $-57,141

For n=6   AW = $-57,300

For n=7: AW = $-58,120

Economic service life is 5 years with AW = $-57,141.

11.15 Marginal costs used to find the ESL of 5 years with AW = $-57,141.

	A	B	C	D	E	F	G	H
1								
2		Market	Loss in MV	Lost interest		MC for	AW of	
3	Year	value	for year	MV for year	AOC	year	marginal cost	
4	0	$100,000						
5	1	$ 75,000	$ 25,000	$ 14,000	$28,000	$ 67,000	$ (67,000)	
6	2	$ 60,000	$ 15,000	$ 10,500	$31,000	$ 56,500	$ (62,093)	
7	3	$ 50,000	$ 10,000	$ 8,400	$34,000	$ 52,400	$ (59,275)	
8	4	$ 40,000	$ 10,000	$ 7,000	$34,000	$ 51,000	$ (57,594)	
9	5	$ 25,000	$ 15,000	$ 5,600	$34,000	$ 54,600	$ (57,141)	ESL
10	6	$ 15,000	$ 10,000	$ 3,500	$45,000	$ 58,500	$ (57,300)	
11	7	$ -	$ 15,000	$ 2,100	$49,000	$ 66,100	$ (58,120)	
12								

11.16 No study period is specified. The defender ESL is 2 years with AW = $-81,000. Retain the defender for 2 years, then replace.

11.17 $AW_D = -(100,000 + 20,000)(A/P,20\%,4) + 40,000(A/F,20\%,4)$
    $= -120,000(0.38629) + 40,000(0.18629)$
    $= \$-38,903$

$AW_C = -300,000(A/P,20\%,10) + 50,000(A/F,20\%,10)$
    $= -300,000(0.23852) + 50,000(0.03852)$
    $= \$-69,630$

Upgrade the existing controlled environment rooms and plan to keep then for 4 years.

11.18 (a) The n values are set; calculate the AW values directly.

$AW_D = -25,000(A/P,20\%,5) - 160,000$
    $= -25,000(0.33438) - 160,000$
    $= \$-168,360$

$AW_C = -700,000(A/P,20\%,10) - 70,000 + 50,000(A/F,20\%,10)$

$$= -700{,}000(0.23852) - 70{,}000 + 50{,}000(0.03852)$$
$$= \$-235{,}038$$

Retain the current bleaching process for 5 more years.

(b) Find the replacement value for the current process.

$$-RV(A/P,20\%,5) - 160{,}000 = -700{,}000(A/P,20\%,10) - 70{,}000 + 50{,}000(A/F,20\%,10)$$

$$-0.33438\ RV = -700{,}000(0.23852) - 70{,}000 + 160{,}000 + 50{,}000(0.03852)$$

$$-0.33438\ RV = -75{,}038$$

$$RV = \$224{,}409$$

This is 50% of the original cost 8 years ago.

11.19 No study period is stated. Find the AW of the defender for 1 year and 2 years. Subscripts are D1 and D2.

$$AW_{D1} = -(3000 + 12{,}000)(A/P,18\%,1) - 22{,}000 + 3000(A/F,18\%,1)$$
$$= -15{,}000(1.1800) - 22{,}000 + 3000(1.000)$$
$$= \$-36{,}700$$

$$AW_{D2} = -(3000 + 12{,}000)(A/P,18\%,2) - 22{,}000 + (-3000 + 3000)(A/F,18\%,2)$$
$$= -15{,}000(0.63872) - 22{,}000$$
$$= \$-31{,}581$$

The defender ESL is 2 years with $AW_D = \$-31{,}581$.

$AW_C = \$-29{,}630$

The company should replace the machine now.

11.20 A new ESL must be determined if the estimates for the defender or challenger have changed during the year significantly enough to make a difference in the ESL and the AW value.

11.21 Find the ESL of the defender over 1, 2 and 3 years. Compare it with the AW of challenger for 10 years.

$$AW_{D1} = -8000(A/P,15\%,1) - 50{,}000 + 6000(A/F,15\%,1)$$
$$= -8000(1.15) - 44{,}000$$
$$= \$-53{,}200$$

$AW_{D2} = -8000(A/P,15\%,2) - 50,000 + (-3000 + 4000)(A/F,15\%,2)$
$\quad = -8000 (0.61512) - 50,000 + 1000(0.46512)$
$\quad = \$-54,456$

$AW_{D3} = -8000(A/P,15\%,3) - [50,000(P/F,15\%,1) + 53,000(P/F,15\%,2)]$
$\quad (A/P,15\%,3) + (-60,000 + 1000)(A/F,15\%,3)$

$\quad = -8000 (0.43798) - [50,000(0.8696) + 53,000(0.7561)] (0.43798)$
$\quad\quad -59,000(0.28798)$

$\quad = -\$57,089$

The ESL is now 1 year with $AW_D = \$-53,200$.

$AW_C = -125,000(A/P,15\%,10) - 31,000 + 10,000(A/F,15\%,10)$
$\quad = -125,000(0.19925) - 31,000 + 10,000(0.04925)$
$\quad = \$-55,414$

The company should replace the machine after 1 year.

11.22 The opportunity cost refers to the money that is foregone by keeping the existing asset.

11.23 The cash flow approach will only yield the proper decision when the defender and challenger have the same lives. Also, the cash flow approach does not properly reflect the amount of money needed to recover the initial investment because the value used for P (i.e., first cost – market value of defender) is lower than it should be.

11.24 Find the replacement value (RV) for the in-place system.

$-RV(A/P,12\%,7) - 27,000 + 50,000(A/F,12\%,7) = -400,000(A/P,12\%,12) -$
$\quad 50,000 + 35,000(A/F,12\%,12)$

$-RV(0.21912) - 27,000 + 50,000(0.09912) = -400,000(0.16144) -$
$\quad 50,000 + 35,000(0.04144)$

$-0.21912\ RV = -91,082$
$\quad RV = \$415,670$

11.25 Perform an ESL analysis for both sites and apply the replacement study steps.

<u>Existing Site</u>
$AW_{D1} = -(180,000 + 200,000)(A/P,20\%,1) + 390,000$
$\quad = -380,000(1.2000) + 390,000$
$\quad = \$-66,000$

$AW_{D2} = -(180,000 + 200,000)(A/P,20\%,2) + 410,000(A/F,20\%,2)$
$= -380,000(0.65455) + 410,000(0.45455)$
$= \$-62,364$

$AW_{D3} = \$-66,385$

$AW_{D4} = \$-66,686$

$AW_{D5} = \$-67,937$

ESL for the defender is 2 years with $AW_D = \$-62,364$.

New Site
$AW_{C1} = -600,000(A/P,20\%,1) + 80,000 + 620,000$
$= \$-20,000$

$AW_{C2} = \$-600,000(A/P,20\%,2) + 80,000 + 10,000(A/G,20\%,2) + 640,000(A/F,20\%,2)$
$= \$-17,273$

$AW_{C3} = \$-600,000((A/P,20\%,3) + 80,000 + 10,000(A/G,20\%,3) + 670,000(A/F,20\%,3)$
$= \$-11,978$

$AW_{C4} = \$-8629$

$AW_{C5} = \$-3438$

ESL for the challenger is 5 years with $AW_C = \$-3438$.

Apply step 1 of the replacement study procedure:

Decision: Replace the defender (present site) with the challenger (new site) now and plan to retain it for 5 years with the lowest equivalent cost of \$3438 per year.

11.26 Determine ESL of defender and challenger and then decide how long to keep defender.

Defender ESL analysis for 1, 2 and 3 years:

$AW_{D1} = -20,000(A/P,15\%,1) - 52,000 + 9000$
$= -20,000(1.15) - 43,000$
$= \$-66,000$

$$AW_{D2} = -20{,}000(A/P,15\%,2) - 52{,}000(P/F,15\%,1)(A/P,15\%,2)$$
$$+ (-61{,}000 + 2000)(A/F,15\%,2)$$
$$= -20{,}000(0.61512) - 52{,}000(0.8696)(0.61512) - 59{,}000(0.46512)$$
$$= \$\text{-}67{,}558$$

$$AW_{D3} = -20{,}000(A/P,15\%,3) - [52{,}000(P/F,15\%,1) + 61{,}000(P/F,15\%,2)]$$
$$(A/P,15\%,3) + (-68{,}000 + 1000)(A/F,15\%,3)$$
$$= -20{,}000(0.43798) - [52{,}000(0.8696) + 61{,}000(0.7561)](0.43798)$$
$$- 67{,}000(0.28798)$$
$$= \$\text{-}68{,}060$$

Year	Market value	AOC	Total AW
0	$20,000	--	--
1	9,000	$-52,000	$-66,000
2	2,000	-61,000	-67,558
3	1,000	-68,000	-68,060

Defender ESL is 1 year.

Challenger ESL analysis for 1 through 6 years:

$$AW_{C1} = -130{,}000(A/P15\%,1) - 30{,}000 + 95{,}000$$
$$= -130{,}000(1.15) + 65{,}000$$
$$= \$\text{-}84{,}500$$

$AW_{C2} = \$\text{-}76{,}012$

$AW_{C3} = \$\text{-}76{,}080$

$AW_{C4} = \$\text{-}72{,}209$

$AW_{C5} = \$\text{-}69{,}990$

$AW_{C6} = \$\text{-}70{,}620$

Challenger ESL is 5 years.

Decision: Keep the defender 1 more year at AW = $-66,000, then replace for 5 years with $AW_C = \$\text{-}69{,}990$.

11.27 (a) Challenger life is n ≤ 6 years. The defender can be retained only 2 or 3 years. For the 8-year study period, the options are:

Options	Defender	Challenger
1	2 years	6 years
2	3	5

$$AW_{D2} = -20{,}000(A/P,15\%,2) - 52{,}000(P/F,15\%,1)(A/P,15\%,2)$$
$$\qquad + (-61{,}000 + 2000)(A/F,15\%,2)$$
$$\qquad = -20{,}000(0.61512) - 52{,}000(0.8696)(0.61512) - 59{,}000(0.46512)$$
$$\qquad = \$\text{-}67{,}558$$

$$AW_{D3} = -20{,}000(A/P,15\%,3) - [52{,}000(P/F,15\%,1) + 61{,}000(P/F,15\%,2)]$$
$$\qquad (A/P,15\%,3) + (-68{,}000 + 1000)(A/F,15\%,3)$$
$$\qquad = -20{,}000(0.43798) - [52{,}000(0.8696) + 61{,}000(0.7561)](0.43798)$$
$$\qquad\quad - 67{,}000(0.28798)$$
$$\qquad = \$\text{-}68{,}060$$

For the challenger:

$$AW_{C5} = -130{,}000(A/P,15\%,5) + 30{,}000(A/F,15\%,5) - [30{,}000(P/F,15\%,1) +$$
$$\qquad 32{,}000(P/F,15\%,2) + \ldots + 47{,}000(P/F,15\%,5)](A/P,15\%,5)$$
$$\qquad = -130{,}000(0.29832) + 30{,}000(0.14832) - [230{,}222](0.29832)$$
$$\qquad = \$\text{-}69{,}990$$

$AW_{C6} = \$\text{-}70{,}620$

Determine the cash flows for 8 years for each option and calculate their PW values.

Option	Retained, years D	C	Cash flow, $/year 1	2	3	4-8	PW, $
1	2	6	-67,558	-67,558	-70,620	-70,620	-311,917
2	3	5	-68,060	-68,060	-68,060	-69,990	-309,661

Select option 2 (smaller PW of costs); retain defender for 3 years then replace with the challenger for 5 years.

(b) The answers are different. The study period limitation to 8 years does not use the three assumptions of a replacement study with no study period. In this problem, the defender services cannot be obtained after its retention period. In the previous problem, they could. (See the beginning of Sections of 11.3 and 11.5.)

11.28 Study period is 3 years. Three options are viable: defender for 2 more years, challenger for 1; defender 1 year, challenger for 2 years; and, challenger for 3 years. Find the AW values and select the best option.

<u>1. Defender 2 years, challenger 1 year:</u>
$AW = -200{,}000 - (300{,}000 - 200{,}000)(A/F,18\%,3)$
$\qquad = -200{,}000 - 100{,}000\,(0.27992)$
$\qquad = \$\text{-}227{,}992$

2. Defender 1 year, challenger 2 years
AW = [-200,000(P/F,18%,1) + 225,000(P/A,18%,2)(P/F,18%,1)](A/P,18%,3)
   =[-200,000(0.8475) + 225,000(1.5656)(0.8475)](0.45992)
   = $-215,261

3. Challenger for 3 years
AW = $-275,000

Decision: Replace the defender after 1 year.

11.29 (a) If no study period is specified, the 3 replacement study assumptions hold. So, the services of the defender and challenger can be obtained (it is assumed) at their AW values. Specifying the study period takes these assumptions away and does not assume repeatability of either D or C alternatives.

(b) If study period is specified, the viable options must be evaluated. Without a study period, the ESL analysis, or the AW values at the set n values, determine the AW values for D and C. Selection of the best option or alternative gives the decision.

11.30 (a)

Option	Defender	Challenger
1	0	5
2	0	6
3	0	7
4	0	8
5	3	2
6	3	3
7	3	4
8	3	5

Find AW values for different options for defender and challenger use periods. Since the AW of the defender ($-70,000 per year) is lower than any of the challenger AW values, the defender should be kept for its remaining life of 3 years. There is no need to evaluate the first 4 options.

The AW subscript is the option number, which is also the study period length.

AW_5 = [-70,000(P/A,15%,3) – 90,000(P/A,15%,2) (P/F,15%,3)](A/P,15%,5)
      = $-76,378

AW_6 = [-70,000(P/A,15%,3) – 80,000(P/A,15%,3) (P/F,15%,3)](A/P,15%,6)
      = $-73,967

AW_7 = [-70,000(P/A,15%,3) – 80,000(P/A,15%,4) (P/F,15%,3)](A/P,15%,7)
      = $-74,512

$$AW_8 = [-70{,}000(P/A,15\%,3) - 80{,}000(P/A,15\%,5)(P/F,15\%,3)](A/P,15\%,8)$$
$$= \$-74{,}912$$

(b) The lowest AW is for option 6 ($-73,967) with a 6-year planning horizon.

11.31 There are only two options: defender for 3, challenger for 2 years; defender for 0, challenger for 5.

Defender

$$AW_{D3} = -(70{,}000 + 40{,}000)(A/P,20\%,3) - 85{,}000$$
$$= -110{,}000(0.47473) - 85{,}000$$
$$= \$-137{,}220$$

Challenger

$$AW_{C2} = -220{,}000(A/P,20\%,2) - 65{,}000 + 50{,}000(A/F,20\%,2)$$
$$= -220{,}000(0.65455) - 65{,}000 + 50{,}000(0.45455)$$
$$= \$-186{,}274$$

$$AW_{C3} = \$-155{,}703$$

$$AW_{C4} = \$-140{,}669$$

$$AW_{C5} = -220{,}000(A/P,20\%,5) - 65{,}000 + 50{,}000(A/F,20\%,5)$$
$$= -220{,}000(0.33438) - 65{,}00 + 50{,}000(0.13438)$$
$$= \$-131{,}845$$

The challenger AW of $-131,845 for 5 years of service is lower than that of the defender. Therefore, by inspection, the defender should be replaced now.

Option 1: defender 3 years, challenger 2 years

$$AW = [-137{,}220(P/A,20\%,3) - 186{,}274(P/A,20\%,2)(P/F, 20\%, 3)](A/P,20\%,5)$$
$$= \$-151{,}726$$

Option 2: defender replaced now, challenger for 5 years

$$AW = \$-131{,}845$$

Again, replace the defender with the challenger now.

FE Review Solutions

11.32   Answer is (d)

11.33   Answer is (b)

11.34   Answer is (c)

11.35   Answer is (b)

Extended Exercise Solutions

The three spreadsheets below answer the three questions.

1. The ESL is 13 years.

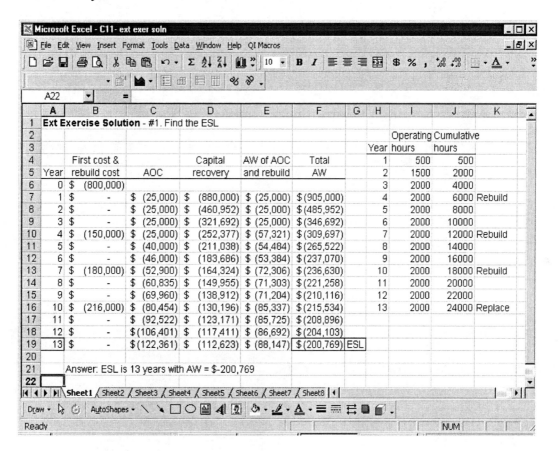

2. Required MV = $1,420,983 found using SOLVER with F12 the target cell and B12 the changing cell.

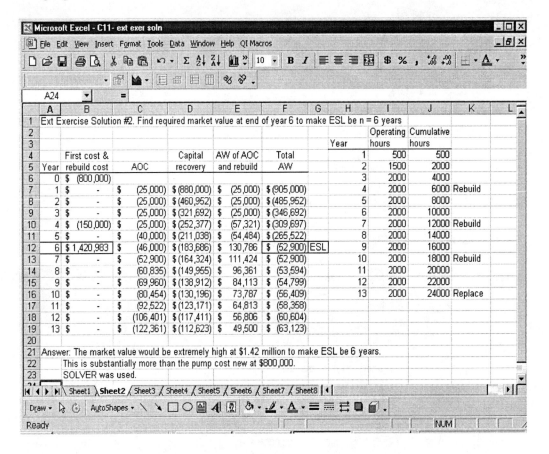

3. SOLVER yields the base AOC = $-201,983 in year 1 with increases of 15% per year. The rebuild cost in year 4 (after 6000 hours) is $150,000. Also this AOC series is huge compared to the estimated AOC of $25,000 (years 1 – 4).

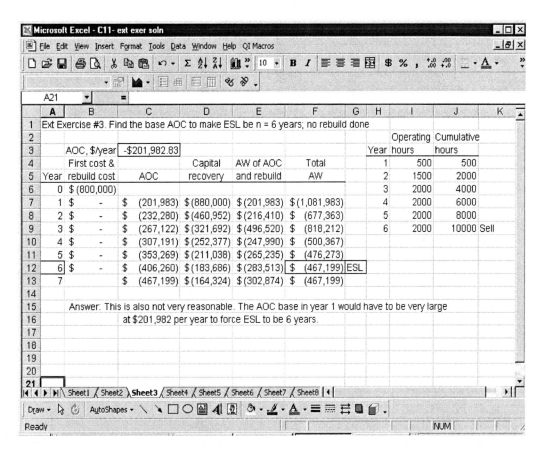

Neither suggestion in #2 or #3 are good options.

Chapter 12 – Selection from Independent Projects Under Budget Limitation
Solutions to end of chapter exercises

Problems

12.1 Three characteristics are: budgetary constraint on total capital invested; select each project entirely or no select it; maximize the return using a measure such as PW.

12.2 There are $2^4 = 16$ possible bundles. Considering the selection restrictions, the 8 viable bundles are:

2	23
3	34
4	134
14	DN

Not acceptable are: 1, 12, 13, 24, 123, 234, 1234.

12.3 There are $2^4 = 16$ possible bundles. Considering the selection restrictions and the $18 million limitation, the 4 viable bundles are:

Projects	Investment
2	$15 million
3	8
2,4	18
3,4	11

Bundles not acceptable due to restrictions or budget over-commitment are:
1, 4, 12, 13, 14, 23, 123, 124, 134, 234, 1234, DN.

12.4 The reinvestment assumption is that any net positive cash flows that occur in any project are reinvested at the MARR from the time they are realized until the end of the longest-lived project being evaluated. (This is similar to the assumption made in Section 7.5 when the composite rate of return is determined, but here the only rate involved is the MARR.) In effect, this makes the lives equal for all projects, a requirement to correctly apply the PW method.

12.5 (a) Develop the bundles with less than $325,000 investment, and select the one with the largest PW value.

Bundle	Projects	Initial investment, $	NCF, $/year	PW at 10%, $
1	A	-100,000	50,000	166,746
2	B	-125,000	24,000	3,038
3	C	-120,000	75,000	280,118
4	D	-220,000	39,000	-11,938
5	E	-200,000	82,000	237,464
6	AB	-225,000	74,000	169,784
7	AC	-220,000	125,000	446,864
8	AD	-320,000	89,000	154,807
9	AE	-300,000	132,000	404,208
10	BC	-245,000	99,000	283,156
11	BE	-325,000	106,000	240,500
12	CE	-320,000	157,000	517,580
13	DN	0	0	0

$PW_1 = -100,000 + 50,000(P/A,10\%,8)$
   $= -100,000 + 50,000(5.3349)$
   $= \$166,746$

$PW_2 = -125,000 + 24,000(P/A,10\%,8)$
   $= -125,000 + 24,000(5.3349)$
   $= \$3038$

$PW_3 = -120,000 + 75,000(P/A,10\%,8)$
   $= -120,000 + 75,000(5.3349)$
   $= \$280,118$

$PW_4 = -220,000 + 39,000(P/A,10\%,8)$
   $= -220,000 + 39,000(5.3349)$
   $= \$-11,939$

$PW_5 = -200,000 + 82,000(P/A,10\%,8)$
   $= -200,000 + 82,000(5.3349)$
   $= \$237,462$

All other PW values are obtained by adding the respective PW for the bundles 1 through 5.

Select PW = $517,580, which is bundle 12 (projects C and E) with $320,000 total investment.

(b) For mutually exclusive projects, select the one project with the largest PW. This is C with PW = $280,118.

Chapter 12

12.6 (a) PW analysis of the 6 viable bundles is shown below. NPV functions used to find PW values. Select projects A and B for a total of $550,000.

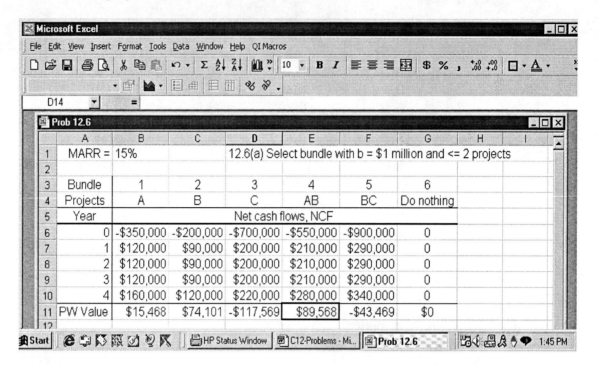

(b) Change the NCF for bundle 5 (B and C) such that the PW is equal to $PW_4 = \$89,568$. Either the NCF can be changed manually, or the SOLVER tool can be used to find the necessary minimum NCF of $336,598, as shown below.

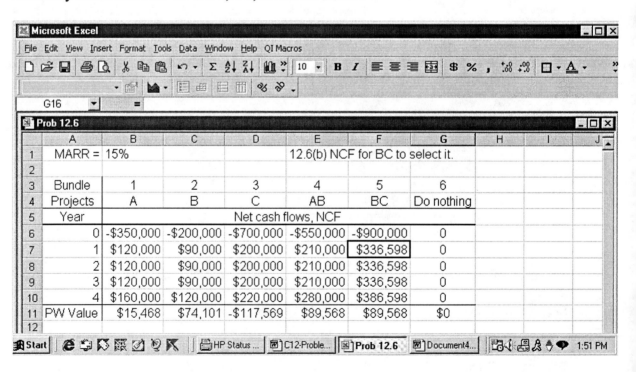

12.7 b = $80,000   i = 15% n_j = 4 years   6 viable bundles

Bundle, j	Projects	NCF_{j0}	NCF_{jt}	S	PW_j at 15%
1	I	$-25,000	$ 6,000	$ 4,000	$-5,582
2	II	-30,000	9,000	-1,000	-4,877
3	III	-50,000	15,000	20,000	4,261
4	I, II	-55,000	15,000	3,000	-10,459*
5	I, III	-75,000	21,000	24,000	- 1,321*
6	II, III	-80,000	24,000	19,000	- 616*

$PW_j = NCF_j(P/A,15\%,4) + S(P/F,15\%,4) - NCF_{j0}$

*Add PW_j values for j = 4, 5 and 6. Since PW_1 and PW_2 < 0 and the loss exceeds the profit in PW_3, by inspection, bundles 4, 5 and 6 will have PW < 0. There is no need to determine their PW values.

Select project III.

12.8 (a) There are 2^7 = 128 separate bundles possible. Only 1, 2 or 3 projects can be accepted. With the budget limitation of $400,000 and selection restrictions, there are only 8 viable bundles.

Bundle	Projects	Initial investment, $	PW at 12%, $
1	2	-200,000	19,000
2	3	- 95,000	39,000
3	5	-195,000	108,000
4	23	-295,000	58,000
5	25	-395,000	127,000
6	27	-300,000	81,000
7	35	-290,000	147,000
8	237	-395,000	120,000

Select projects 3 and 5 with PW = $147,000 and $290,000 invested.

(b) The second good choice is projects 2 and 5 with PW = $127,000. It invests $395,000.

12.9 (a) At b = $30,000 only 5 bundles are viable of the 32 possible ones.

Bundle	Projects	Initial investment, $	PW at 18%, $
1	T	-15,000	8,540
2	R	-25,000	12,325
3	U	-10,000	3,005
4	M	-25,000	10
5	TU	-25,000	11,545

Select project R with PW = $12,325 and $25,000 invested.

(b) With b = $60,000, 11 more bundles are viable.

Bundle	Projects	Initial investment, $	PW at 18%, $
6	P	-40,000	15,350
7	TR	-40,000	20,865
8	TM	-40,000	8,550
9	TP	-55,000	23,890
10	RU	-35,000	15,330
11	RM	-50,000	12,335
12	UM	-35,000	3,015
13	UP	-50,000	18,355
14	TRU	-50,000	23,870
15	TUM	-50,000	11,555
16	RUM	-60,000	15,340

Select projects T and P with PW = $23,890 and $55,000 invested.
(A close second are projects T, R and U with PW = $23,870 and $50,000 invested.)

(c) Select all projects since they each have PW > 0 at 18%.

12.10 Form all bundles that include 3 projects (less than $125,000) that follow the selection restrictions. These are MNP, MOP, and MPQ. Add the PW values and select the largest one.

Bundle	Projects	PW at 1.5% per month
1	MNP	$3500 + 19,300 + 2500 = $25,300
2	MOP	3500 + 18,200 + 2500 = $24,200
3	MPQ	3500 + 2500 + 22,000 = $28,000

Invest in projects M, P, and Q for $120,000 since its PW is largest.

12.11 (a) The bundles and PW values are determined using the factors at MARR = 15% per year.

Bundle	Projects	Initial investment, $	NCF, $/year	Life, years	PW at 15%
1	1	-1.5 mil	360,000	8	$115,428
2	2	-3.0	600,000	10	11,280
3	3	-1.8	520,000	5	-56,856
4	4	-2.0	820,000	4	341,100
5	13	-3.3	880,000	1-5	58,572
			360,000	6-8	
6	14	-3.5	1,180,000	1-4	456,528
			360,000	5-8	
7	34	-3.8	1,340,000	1-4	284,244
			520,000	5	

Select PW = $456,528 for projects 1 and 4 with $3.5 million invested.

(b) Set up a spreadsheet for all 7 bundles. Select projects 1 and 4 with the largest PW = $456,518 and invest $3,5 million.

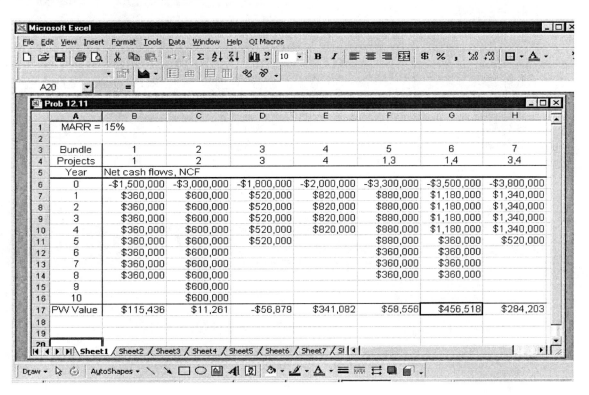

12.12 Spreadsheet below shows the solution. Select project 2 for a budget of $3.0 million and a PW of $328,849.

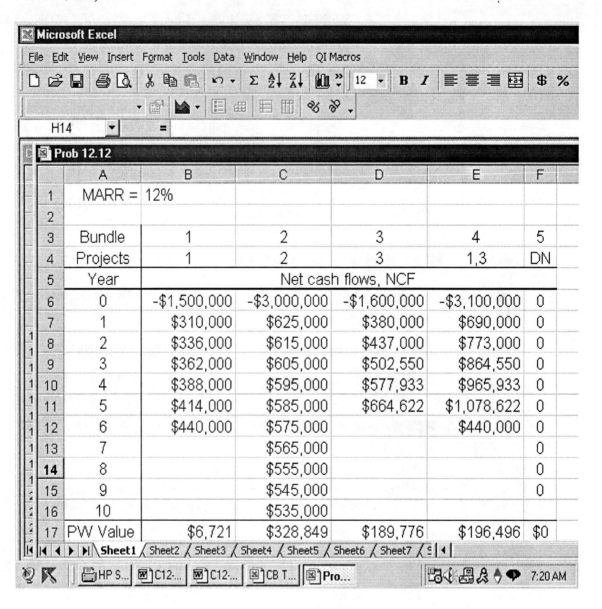

12.13 Budget limit = $16,000   MARR = 12% per year

Bundle	Projects	Investment	NCF for years 1 through 5	PW at 12%
1	1	$-5,000	$1000,1700,2400, 3000,3800	$3019
2	2	-8,000	500,500,500, 500,10500	-523
3	3	-9,000	5000,5000,2000	874
4	4	-10,000	0,0,0,17000	804
5	1,2	-13,000	1500,2200,2900, 3500,14300	2496
6	1,3	-14,000	6000,6700,4400, 3000,3800	3893
7	1,4	-15,000	1000,1700,2400, 20000,3800	3823

Since PW_6 = $3893 is largest, select bundle 6, which is projects 1 and 3.

12.14 For the bundle comprised of projects 3 and 4 the net cash flows are:

Year	0	1	2	3	4	5
NCF	$-19,000	5000	5000	2000	17,000	0

Use Eq. [12.2] to compute the PW value at 12%. The longest-life project of the four is n_L = 5 years.

$$PW = -19,000 + [5,000(F/A,12\%,2)(F/P,12\%,3) + 2,000(F/P,12\%,2) + 17,000(F/P,12\%,1)] (P/F,12\%,5)$$

$$= -19,000 + [5,000(2.12)(1.4049) + 2,000(1.2544) + 17,000(1.12)] (0.5674)$$
$$= \$1676$$

The PW values using the NCF values directly is

$$PW = -19,000 + 5000(P/A,12\%,2) + 2000(P/F,12\%,3) + 17,000(P/F,12\%,4)$$
$$= -19,000 + 5000(1.6901) + 2000(0.7118) + 17,000(0.6355)$$
$$= \$1677$$

The PW values are the same (allowing for round-off error to the nearest dollar).

12.15 To complete the ILP formulation, find the PW of project E.

$$PW_E = -21{,}000 + 9500(P/A,15\%,9)$$
$$= -21{,}000 + 9500(4.7716)$$
$$= \$24{,}330$$

The linear programming formulation is:

Maximize $Z = 3694x_1 - 1019 x_2 + 4788 x_3 + 6120 x_4 + 24{,}330 x_5$

Constraints: $10{,}000x_1 + 15{,}000 x_2 + 8000 x_3 + 6000 x_4 + 21{,}000 x_5 < 20{,}000$

$x_k = 0$ or 1 for $k = 1$ to 5

The spreadsheet solution using the template in Figure 12-3 is below. The MARR is set to 15% and the SOLVER budget constraint is set to $20,000. Projects C and D are selected (row 19) for a $14,000 investment (cell I22) with Z = $10,908 (cell I5), as in Example 12.1.

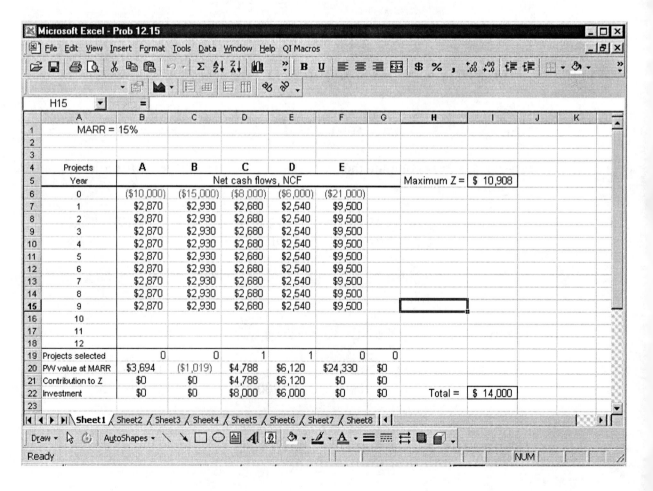

Chapter 12

12.16 Use the capital budgeting template with MARR = 10% and the budget constraint at $325,000. The solution is to select projects C and E (row 19) with $320,000 invested and a maximized PW = $517,583 (cell I5).

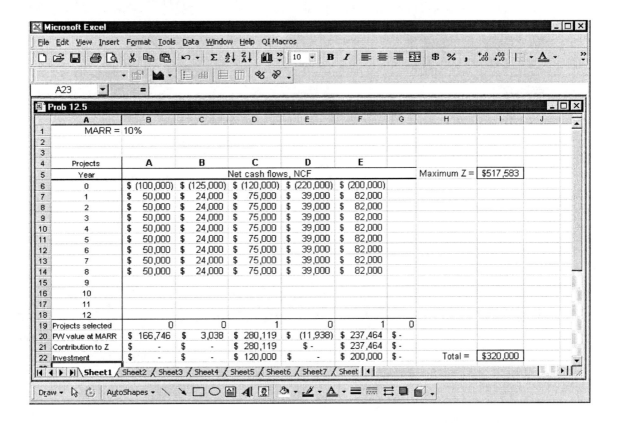

12.17 The capital budgeting solution is below with MARR = 15 and b = $80,000. Select project III with Z = $4260; same as in problem 12.7.

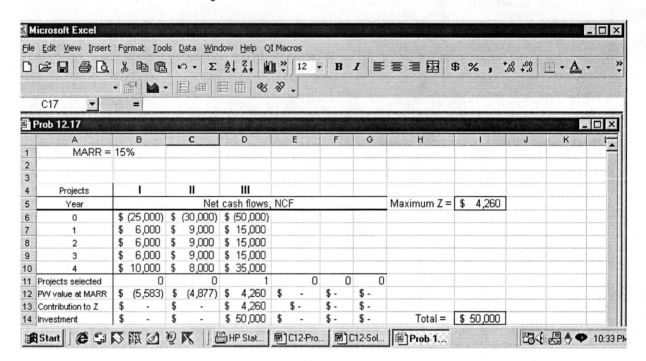

12.18 Set up the spreadsheet with the investment amounts in row 6. Directly enter the PW values in the row for 'PW value at MARR' since there are no annual NCF estimates provided. Then, simply change the constraint for the budget to $30,000 and then to $60,000. For part (c) place no constraint on the budget cell. The answers are the same as Problem 12.9, namely:

(a) Select R with Z = $12,325 and $25,000 invested.
(b) Select T and P with Z = $23,890 and $55,000 invested (see spreadsheet below).
(c) Select all five projects with Z = $39,230 and $115,000 invested.

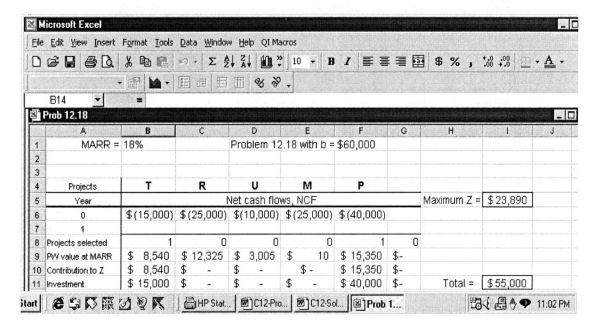

12.19 Use the capital budgeting problem template at 15% and a constraint on cell I22 of $4,000,000. Solution is to select projects 1 and 4 with $3.5 million invested and Z = $456,518.

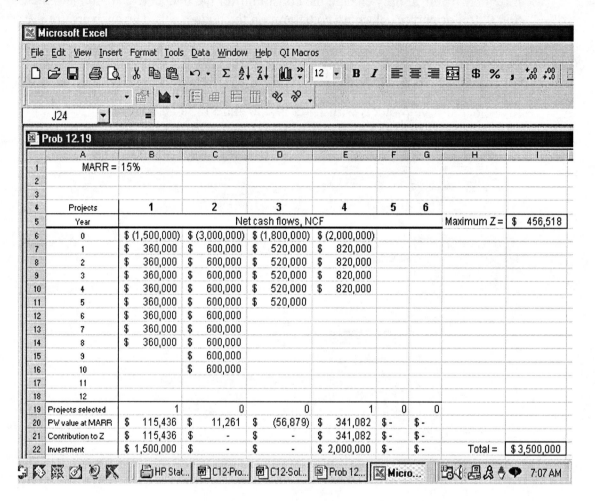

12.20 Enter the NCF values from Problem 12.12 into the capital budgeting template and b = $3,300,000 into SOLVER. Select project 2 for Z = $328,849 with $3.0 million invested.

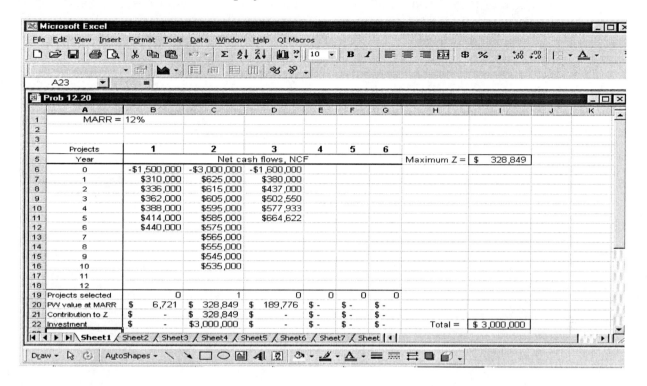

12.21 Enter the four project NCF values, MARR = 12% and b = $16,000 constraint in SOLVER to obtain the answer: Select projects 1 and 3 with Z = $3893 and $14,000 invested, the same as in Problem 12.13 where all viable mutually exclusive bundles were evaluated by hand.

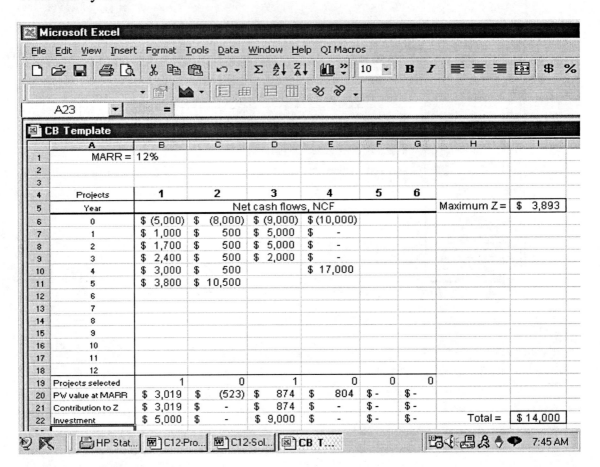

Case Study Solution

(1) Rows 5 and 6 of the spreadsheet show the viable bundles for the $3.5 million spending limit and the project relationship.

(2) Projects B and C with PW = $895,000 are the economic choices. This commits only $2.2 million of the allowed $3.5 million.

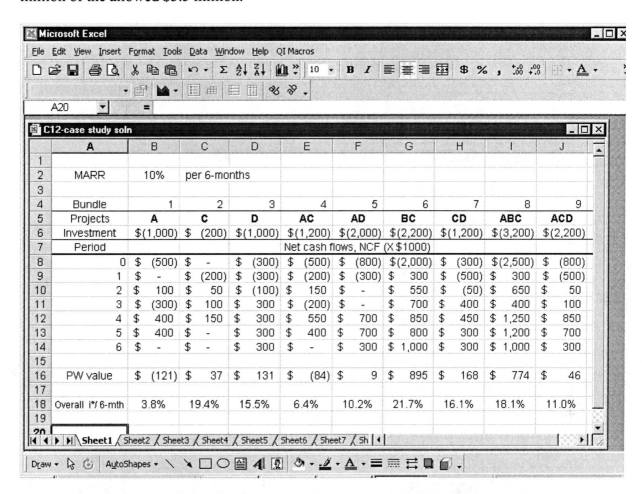

(3) Change cash flows, investment amount, life, etc. to obtain a PW and overall i* greater than the results for BC (column G).

Chapter 13 – Breakeven Analysis
Solutions to end of chapter exercises

Problems

13.1 (a) $Q_{BE} = 1,000,000/(8.90-4.50) = 227,272$ units

    (b)   Profit = R – TC
               $= 8.90Q - 1,000,000 - 4.5Q$

         at 200,000 units: Profit = $8.90(200,000) - 1,000,000 - 4.50(200,000)$
                                  = $-120,000 (loss)

         at 450,000 units: Profit = $980,000

13.2 $15,000,000 (A/P, 1\%, 120) + (4,100,000) E^{1.8} = 12,000 (E)(250)$

       $15,000,000(0.01435) + (4,100,000)E^{1.8} = 3,000,000(E)$

       $215,250 = 3,000,000(E) - 4,100,000E^{1.8}$

          Solve for E by trial and error:

     at E = 0.55:   215,250 < 252,227
     at E = 0.57:   215,250 < 219,409
     at E = 0.58:   215,250 > 202,007

          E = 0.572 or 57.2% minimum removal efficiency

13.3 (a)   AC = TC/Q = (FC + vQ)/Q
                   = FC/Q + v
                   = 60,000/Q + 2

     For AC = $3, find Q

         3 = 60,000/Q + 2
         Q = 60,000/1 = 60,000 units

     Plot AC and estimate Q = 60,000 units for AC = $3

    (b)   AC = 3.50 = 100,000/Q + 2

          Q = 100,000/1.50 = 66,667 units

13.4 (a) $Q_{BE} = \dfrac{775,000}{5-3} = 387,500$ calls per year

This is 96.9% of the center capacity

(b) Set $Q_{BE} = 100,000$ and determine r at v = $3.

$$100,000 = \dfrac{450,000}{r-3}$$

$$r - 3 = \dfrac{450,000}{100,000}$$

$$r = 4.5 + 3 = \$7.50 \text{ per call}$$

Revenue per call required is $2.50 higher.

13.5 Let x = gradient increase per year. Set revenue = cost.

$[4000 + x\,(A/G,12\%,3)](33,000 - 21,000) = -200,000,000\,(A/P,12\%,3) + (0.20)(200,000,000)\,(A/F,12\%,3)$

$[4000 + x\,(0.9246)](12,000) = -200,000,000(0.41635) + 40,000,000(0.29635)$

$x = 2110$ cars/year increase

13.6 (a) Profit = R - TC = $25Q - 0.001Q^2 - 3Q - 2$
        = $-0.001Q^2 + 22Q - 2$

Q	Profit (approximate)
5,000	$ 85,000
10,000	120,000
11,000	121,000
15,000	105,000
20,000	40,000
25,000	-75,000

About 11,000 cases per year is breakeven with profit of $121,000.

(b) In general Profit = R - TC = $aQ^2 + bQ + c$

The a, b and c are constants. Take the first derivative, set equal to 0, and solve.

$Q_{max} = -b/2a$

Chapter 13

Substitute into the profit relation.

$$\text{Profit}_{max} = (-b^2/4a) + c$$

Here, $Q_{max} = 22/2(0.001)$
= 11,000 cases per year
$$\text{Profit}_{max} = [-(22)^2/4(-0.001)] - 2$$
= $120,998 per year

13.7 FC = $305,000   v = $5500/unit

(a)   Profit = (r – v)Q – FC

0 = (r – 5500)5000 – 305,000
(r – 5500) = 305,000 / 5000
r = 61 + 5500
 = $5561 per unit

(b)   Profit = (r – v)Q – FC

500,000 = (r – 5500)8000 – 305,000
(r – 5500) = (500,000 + 305,000) / 8000
r = $5601 per unit

13.8 Let x = ads per year

–12,000(A/P,10%,3) – 45,000 + 2000(A/F,10%,3) – 4x = –21x
–12,000(0.40211) – 45,000 + 2000(0.30211) = –17x

x = 2,895 ads per year

13.9 Let n = number of months

–15,000(A/P, 0.5%, n) – 40 = –1,000
–15,000(A/P, 0.5%, n) = –960
(A/P, 0.5%, n) = 0.064

n is between 16 and 17
n = 17 months

13.10 Let x = number of miles per year

$$-900{,}000(A/P,7\%,20) - 200x = -x[(28)2]/0.1$$

$$-900{,}000(0.09439) - 200x = -560x$$
$$x = 236 \text{ miles per year}$$

13.11 Let x = hours per year

$$-800(A/P,10\%,3) - (300/2000)x - 1.0x = -1{,}900(A/P,10\%,5) - (700/8000)x - 1.0x$$
$$-800(0.40211) - 0.15x - 1.0x = -1{,}900(0.2638) - 0.0875x - 1.0x$$

$$0.0625x = 179.532$$
$$x = 2873 \text{ hours per year}$$

13.12 Set $AW_A = AW_B$, with P_B = first cost of Proposal B. The final term in AW_B removes the repainting cost in year 16.

$$-250{,}000(A/P,12\%,4) - 3{,}000 = -P_B(A/P,12\%,16) - 5{,}000(A/F,12\%,2) + 5{,}000(A/F,12\%,16)$$

$$-250{,}000(0.32923) - 3{,}000 = -P_B(0.14339) - 5{,}000(0.4717) + 5{,}000(0.02339)$$

$$-85{,}307.50 = -P_B(0.14339) - 2241.55$$
$$83{,}065.95 = P_B(0.14339)$$
$$P_B = \$579{,}300$$

13.13 Let x = production in year 3

$$-40{,}000 - 50x = -70{,}000 - 12x$$
$$38x = 30{,}000$$
$$x = 790 \text{ units}$$

13.14 Let FC_B = fixed cost for B. Set total cost relations equal at 2000 units per year.

Variable cost for B = 2000/200 = $10/unit

$$40{,}000 + 60(2000 \text{ units}) = FC_B + 10(2000 \text{ units})$$

$$FC_B = \$140{,}000 \text{ per year}$$

13.15 Let x = breakeven days per year

$$-125{,}000(A/P,12\%,8) + 5{,}000(A/F,12\%,8) - 2{,}000 - 40x = -45(125 + 20x)$$
$$-125{,}000(0.2013) + 5{,}000(0.0813) - 2{,}000 - 40x = -5625 - 900x$$
$$-21{,}131 = -860x$$
$$x = 24.6 \text{ days per year}$$

13.16 Let x = yards per year to breakeven

(a) Solution by hand

$$-40{,}000(A/P,8\%,10) - 2{,}000 - (30/2500)x = -[6(14)/2500]x$$
$$-40{,}000(0.14903) - 2{,}000 - 0.012x = -0.0336x$$
$$-7961.20 = -0.0216x$$
$$x = 368{,}574 \text{ yards per year}$$

(b) Solution by computer

There are many set-ups to work the problem. One is: Enter the parameters for each alternative. Use SOLVER to force the breakeven equation (target cell D16) to equal 0, with a constraint in SOLVER for the total yardage to be the same (cell B8 = C8 here).

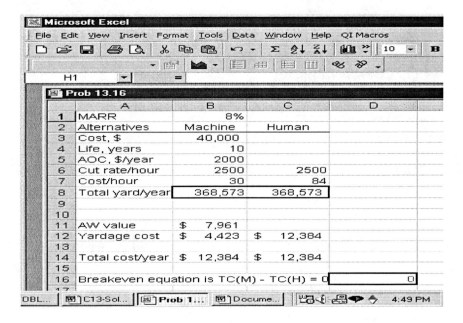

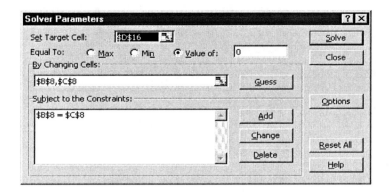

13.17 (a) Let n = number of years. Develop the relation

$$AW_{own} + AW_{lease} + AW_{sell} = 0$$

$-(100,000 + 12,000)(A/P,8\%,n) - 3800 - 2500 - [31000(P/F,8\%,k)](A/P,8\%,n)$
$\quad + 12,000 + (60 + 1.5n)(2,500)(A/F,8\%,n) = 0$

where k = 6, 12, 18, ..., and k ≤ n.

Use trial and error to determine the breakeven n value.

n = 14: $-112,000(0.12130) + 5700 - [1000(0.6302 + 0.3971)](0.12130) +$
$\quad [60 + 1.5(14)](2,500)(0.04130) \stackrel{?}{=} 0$

$-13,586 + 5700 - 125 + 8363 = \$+352 > 0$

n = 16: $-112,000(0.11298) + 5700 - [1000(0.6302 + 0.3971)](0.11298) +$
$\quad [60 + 1.5(16)](2,500)(0.03298) \stackrel{?}{=} 0$

$-12,654 + 5700 - 116 + 6926 = \$-144 < 0$

By interpolation, n = 15.42 years

(b) Selling price = [60 + 1.5(15.42)] (2,500)
= $207,825

13.18 Let x = number of samples per year. Set AW values for complete and partial labs equal to the outsource cost.

(a) Complete lab option
-50,000(A/P,10%,6) - 26,000 - 10x = -120x
-50,000(0.22961) - 26,000 = -110x
x = 341 samples per year

(b) Partial lab option
-35,000(A/P,10%,6) - 10,000 - 3x - 40x = -120x
-35,000(0.22961) - 10,000 = -77x
x = 234 samples per year

(c) Equate AW of complete and partial labs
-50,000(A/P,10%,6) - 26,000 - 10x = -35,000(A/P,10%,6) - 10,000 - 3x - 40x
-50,000(0.22961) - 26,000 - 10x = -35,000(0.22961) - 10,000 − 43x
33x = 19,444
x = 589 samples per year

Ranges for the lowest total cost are:

$0 < x \leq 234$ select outsource
$234 < x \leq 589$ select partial lab
$589 < x$ select complete lab

(d) At 300 samples per year, the partial lab option is the best economically at TC = $30,936.

13.19 Let P = first cost of plastic lining

(a) by hand: $-2,000(A/P,5\%,6) - 600(P/F,5\%,4)(A/P,5\%,6) = -P(A/P,5\%,15)$
$-2,000(0.19702) - 600(0.8227)(0.19702) = -P(0.09634)$
$-491.29 = -P(0.09634)$
$P = \$5,100$

(b) by computer:

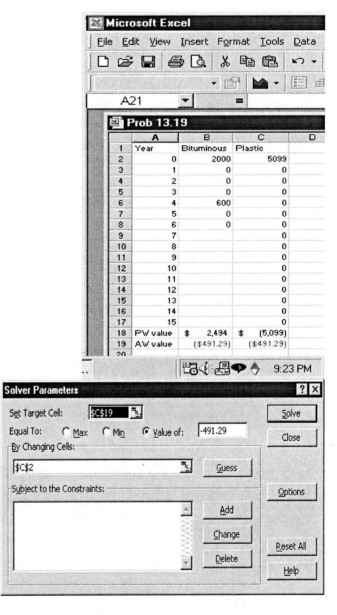

13.20 Let x = savings per year
$-3000(A/P,8\%,10) + 150 = -4600(A/P,8\%,10) + x$
$-3000(0.14903) + 150 = -4600(0.14903) + x$
$-297.09 = -685.54 + x$
$x = \$388.45$ per year

13.21 (a) Let x = days per year to pump the lagoon. Set the AW relations equal.

$$-800(A/P,10\%,8) - 300x = -1600(A/P,10\%,10) - 3x - 12(8200)(A/P,10\%,10)$$
$$-800(0.18744) - 300x = -1600(0.16275) - 3x - 98{,}400(0.16275)$$
$$-149.95 - 300x = -16275 - 3x$$
$$297x = 16125.05$$
$$x = 54.3 \text{ days per year}$$

(b) If the lagoon is pumped 52 times per year and P = cost of pipeline, the breakeven equation in (a) becomes:

$$-800(0.18744) - 300(52) = -1600(0.16275) - 3(52) + P(0.16275)$$
$$-15{,}750 = -416.4 + 0.16275P$$
$$P = \$\text{-}94{,}216$$

13.22 (a) By hand: Let P = first cost of sandblasting. Equate the PW of painting each 4 years to PW of sandblasting each 10 years, up to a total of 38 years for each option.

PW of painting
$$PW_p = -2{,}800 - 3{,}360(P/F,10\%,4) - 4{,}032(P/F,10\%,8) - 4{,}838(P/F,10\%12) - 5{,}806(P/F,10\%,16) - 6{,}967(P/F,10\%,20) - 8{,}361(P/F,10\%,24) - 10{,}033(P/F,10\%,28) - 12{,}039(P/F,10\%,32) - 14{,}447(P/F,10\%,36)$$

$$PW_p = -2{,}800 - 3{,}360(0.6830) - 4{,}032(0.4665) - 4{,}838(0.3186) - 5{,}806(0.2176) - 6{,}967(0.1486) - 8{,}361(0.1015) - 10{,}033(0.0693) - 12{,}039(0.0474) - 14{,}447(0.0323)$$

$$PW_p = \$\text{-}13{,}397$$

PW of sandblasting
$$PW_s = -P - 1.4P(P/F,10\%,10) - 1.96P(P/F,10\%,20) - 2.74P(P/F,10\%,30)$$
$$-P[1 + 1.4(0.3855) + 1.96(0.1486) + 2.74(0.0573)]$$

$$PW_s = -1.988P$$

Equate the PW relations.

$$-13{,}397 = -1.988P$$
$$P = \$6{,}739$$

(b) By computer: Enter the periodic costs. Enter 0 for the P of the sandblasting option. Use SOLVER to find the breakeven P = $6739.

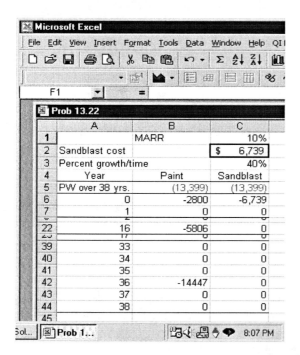

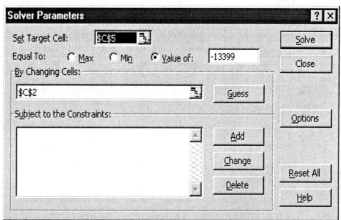

(c) Change cell C3 to 25% and re-SOLVER to get $7337.

Case Study Solution

1. Savings = 40 hp * 0.75 kw/hp * 0.06 $/kwh * 24 hr/day * 30.5 days/mo ÷ 0.90
   = $1464/month

2. A decrease in the efficiency of the aerator motor renders the selected alternative of "sludge recirculation only" *more* attractive, because the cost of aeration would be higher, and, therefore the net savings from its discontinuation would be greater.

3. If the cost of lime increased by 50%, the lime costs for "sludge recirculation only" and "neither aeration nor sludge recirculation" would increase by 50% to $393 and $2070, respectively. Therefore, the cost difference would *increase*.

4. If the efficiency of the sludge recirculation pump decreased from 90% to 70%, the net savings between alternatives 3 and 4 would *decrease*. This is because the $262 saved by not recirculating with a 90% efficient pump would increase to a monthly savings of $336 by not recirculating with a 70% efficient pump.

5. If hardness removal were discontinued, the extra cost for its removal (column 4 in Table 13-1) would be zero for all alternatives. The favored alternative under this scenario would be alternative 4 (neither aeration nor sludge recirculation) with a total savings of $2,471 – 469 = $2002 per month.

6. If the cost of electricity decreased to 4¢/kwh, the aeration and sludge recirculation monthly costs would be $976 and $122, respectively. The net savings for alternative 2 would then be $-1727, for alternative 3 would be $-131, and for alternative four - $751---all losses. Therefore, the best alternative would be number 1, continuation of the normal operating condition.

7. (a) For alternatives 1 and 2 to breakeven, the total savings would have to be equal to the total extra cost of $1,849. Thus,

   1,849/ 30.5 = (5)(0.75)(x)(24) / 0.90
   x = 60.6 cents per kwh

   (b)   1107/ 30.5 = (40)(0.75)(x)(24) / 0.90
   x = 4.5 cents per kwh

   (c)   1,849/ 30.5 = (5)(0.75)(x)(24) / 0.90 + (40)(0.75)(x)(24) / 0.90
   x = 6.7 cents per kwh

Chapter 14 – Effects of Inflation
Solutions to end of chapter exercises

Problems

14.1 There is no difference between today's dollars and constant value dollars.

14.2 Note: constant-value in year t is indicated by CV_t.
$CV_0 = 10,000/(1 + 0.07)^{10}$
$= \$5,083.49$

14.3 (a) $CV_1 = 2000/(1 + 0.06)$
$= \$1,886.79$

$CV_2 = 2000/(1 + 0.06)^2$
$= \$1,779.99$

$CV_3 = 2000/(1 + 0.06)^3$
$= \$1,679.24$

$CV_4 = 2000/(1 + 0.06)^4$
$= \$1,584.18$

$CV_5 = 2000/(1 + 0.06)^5$
$= \$1,494.52$

(b) Sum the CV values to get $8424.72 or use the P/A factor at f = 6%.
P = 2000(P/A,6%,5)
= 2000(4.2124)
= $8424.80

14.4 $CV_5 = 1000 (1 + 0.05)^5$
$= \$1,276.28$

14.5 Next year: Price = $20,000 (1 + 0.029)^1$
= $20,580
3 years: Price = $20,000(1 + 0.029)^3$
= $21,791

14.6 (a) At a 58% increase, $1 would increase to $1.58. Let x = annual increase.
$1.58 = (1 + x)^5$
$(1 + x) = 1.58^{0.2} = 1.0958$
x = 9.58% per year

(b) Amount greater than f = 3% is:
9.58 – 3 = 6.58% per year

14.7 $30{,}000 = 25{,}000(1+f)^6$
$1.2 = (1+f)^6$
$f = 3.09\%$ per year

14.8 (a) Increase each cash flow (CF) by (1.045) per year.
$CF = 1000(1.045)^n$

(b) Decrease each cash flow by 1/1.045 per year.
$CF = 1000(1/1.045)^n$

Year	CF for part (a)	CF for part (b)
1	$1045.00	$956.94
2	1092.03	915.73
3	1141.17	876.30
4	1192.52	838.56
5	1246.18	802.45

14.9 The market interest rate would be equal to the real interest rate when the inflation rate is zero.

14.10 $i_f = 0.03 + 0.04 + (0.03)(0.04)$
$= 7.12\%$ per year

14.11 $0.13 = 0.05 + f + 0.05(f)$
$1.05f = 0.08$
$f = 7.62\%$ per year

14.12 $0.25 = i + 0.18 + i(0.18)$
$0.07 = 1.18i$
$i = 5.93\%$ per year

14.13 (a) $i_f = 0.05 + 0.08 + (0.05)(0.08)$
$= 13.4\%$ per year

$$P = 20{,}000\left[\frac{1}{(1+0.134)^{10}}\right]$$

$= 20{,}000[0.28436]$
$= \$5{,}687$

(b) Use $i = 5\%$.
$P = 20{,}000(P/F, 5\%, 10)$
$= 20{,}000(0.6139)$
$= \$12{,}278$

14.14 (a) $F = 100,000 (1 + 0.04)^{20}$
    $= \$219,112$
    Amount of increase is \$119,112.

(b) $F = 100,000(1.1)^{20}$
    $= \$672,750$
    Amount of increase $= \$572,750$.

(c) $0.10 = i + 0.04 + i(0.04)$
    $1.04i = 0.06$
    $i = 0.0577$
    $= 5.77\%$
    Amount over safe investment is:
    $5.77 - 3.5 = 2.27\%$ per year

14.15 (a) $P = 30,000(P/F,10\%,5)$
    $= 30,000(0.6209)$
    $= \$18,627$

Buy now; the \$15,000 is lower equivalent cost now.

(b) $i_f = 0.10 + 0.06 + 0.10 (0.06)$
    $= 16.6\%$

$$P = 30,000 \left[\frac{1}{(1 + 0.166)^5} \right]$$

    $= 30,000 [0.46399]$
    $= \$13,920$

Buy later; the \$15,000 is a higher equivalent cost now.

14.16 (a) $PW_A = -31,000 - 28,000(P/A,6\%,6) + 5000(P/F,6\%,6)$
    $= -31,000 - 28,000(4.9173) + 5000(0.7050)$
    $= \$-165,159$

$PW_B = -48,000 - 19,000(P/A,6\%,6) + 7000(P/F,6\%,6)$
    $= -48,000 - 19,000(4.9173) + 7000(0.7050)$
    $= \$-136,494$

Purchase Machine B.

(b) $i_f = 0.06 + 0.03 + 0.06(0.03)$
    $= 9.18\%$
    $PW_A = -31,000 - 28,000(P/A,9.18\%,6) + 5000(P/F,9.18\%,6)$
    $= \$-152,814$

$$PW_B = -48,000 - 19,000(P/A,9.18\%,6) + 7000(P/F,9.18\%,6)$$
$$= \$-128,409$$

Purchase Machine B.

14.17 $i_f = 0.14 + 0.02 + (0.14)(0.02)$
= 16.28% per year

For alternative C, first find AW and then divide by i_f.

$AW_C = -8,500,000(A/P,16.28\%,5) - 8000 + 5000(A/F,16.28\%,5)$
= $-2,620,262

Cap. Cost$_C$ = $\dfrac{2,620,262}{0.1628}$
= $-16,094,975

Cap. Cost$_D$ = $-20,000,000 - \dfrac{7000}{0.1628}$
= $-20,042,998

Select alternative C.

14.18 Use the real rate of return for salesman A and the inflated rate of return for B estimates:

$i_f = 0.10 + 0.06 + 0.10(0.06)$
= 16.6%

$PW_A = -60,000 - 25,000(P/A,10\%,10)$
= $-60,000 - 25,000(6.1446)$
= $-\$213,615$

$PW_B = -95,000 - 35,000(P/A,16.6\%,10)$
= $-\$260,452$

Recommend purchase from salesman A.

14.19 (a) Case 1: $F = 10,000(F/P,12\%,6)$
= 10,000(1.9738)
= $19,738

(b) Case 2: Buying power = $\dfrac{19,738}{(1+f)^6}$
= $\dfrac{19,738}{(1+0.05)^6}$
= $14,729

(c) $0.12 = i + 0.05 + i(0.05)$
    $0.07 = 1.05i$
    $i = 6.67\%$ per year

14.20 $F = 25,000(F/P,10\%,4)$
       $= 25,000(1.4641)$
       $= \$36,603$

Buying power $= \dfrac{36,603}{(1+0.06)^4}$
$= \$28,993$

This is case 2.

14.21 (a) Case 4. Find $MARR_f$.
       $MARR_f = 0.12 + 0.05 + 0.12(0.05)$
       $= 17.6\%$

(b) $F = 20,000(F/P,17.6\%,3)$
    $= 20,000(1 + 0.176)^3$
    $= 20,000(1.62638)$
    $= \$32,528$

14.22 Price $= 40,000(1 - 0.02)^3$
       $= \$37,648$

14.23 Price $= 100,000(1 - 0.02)^{10}$
       $= \$81,707$

14.24 First find the rate of return i_f and then solve for the real rate, i.

$2,500,000 = 1,000,000(F/P,i_f,6)$
$2.5 = (1 + i_f)^6$
$i_f = 16.5\%$ per year

$0.165 = i + 0.05 + i(0.05)$
$0.115 = 1.05i$
$i = 10.95\%$ per year

The real MARR is 12%. Since 10.95% < 12%, the investment is not economically justified.

14.25 (a) $653,000 = 150,000(F/P,f,95)$
       $4.3533 = (1 + f)^{95}$
       $f = 1.56\%$ per year

(b) $F = 653,000(1 + 0.03)^{14}$
  $= 653,000(1.5126)$
  $= \$987,728$

(c) Case 3.

14.26 $F = P[(1 + i)(1 + f)(1 + g)]^n$
  $= 250,000[(1.04)(1.035)(1.02)]^5$
  $= 250,000[1.097928]^5$
  $= 250,000(1.5954)$
  $= \$398,850$

14.27 (a) In constant-value dollars, use i = 10% to recover the investment:

  $A = 40$ million $(A/P,10\%,10)$
   $= 40$ million (0.16275)
   $= \$6,510,000$ per year

(b) In future dollars, use i_f to recover the investment:

  $i_f = 0.10 + 0.04 + 0.10(0.04)$
   $= 14.4\%$ per year

  $A = 40$ million $(A/P,14.4\%,10)$
   $= 40$ million (0.19472)
   $= \$7,788,617$ per year   (PMT function)

14.28 $i_f = 0.10 + 0.05 + 0.10(0.05)$
   $= 15.5\%$ per year

  $A = 750,000(A/P,15.5\%,5)$
   $= \$226,391$ (PMT function)

14.29 Find the amount needed in year 10 at 4% inflation.
  Future amount $= 5000(1 + 0.04)^{10} = \$7401$

  Now use 9% to find A.

  $A = F(A/F,9\%,10)$
   $= 7401(0.06582)$
   $= \$487$ per year

14.30 (a) To maintain purchasing power:

  Future amount $= 5$ million $(F/P,6\%,4)$
         $= \$6,312,500$

(b) A = 6,312,500(A/F,10%,4)
 = 6,312,500(0.21547)
 = $1,360,154 per year

14.31 i_f = 0.08 + 0.04 + 0.08(0.04)
 = 12.32% per year

First find P and multiply by i_f:

P = 50,000 + 10,000(P/F,12.32%,5) + $\frac{5000}{0.1232}$ (P/F,12.32%,5)

 = 50,000 + 10,000(0.55939) + 40,584(0.55939)
 = $78,296

A = 78,296(0.1232)
 = $9,646 per year

14.32 (a) Use MARR = 12% per year
AW_A = -80,000(A/P,12%,5) – 70,000 + 10,000(A/F,12%,5)
 = -80,000(0.27741) – 70,000 + 10,000(0.15741)
 = $-90,619

AW_B = -125,000(A/P,12%,5) – 50,000 + 20,000(A/F,12%,5)
 = -125,000(0.27741) – 50,000 + 20,000(0.15741)
 = $-81,528

Select Machine B.

(b) Find and use $MARR_f$
$MARR_f$ = 0.12 + 0.07 + 0.12(0.07)
 = 19.84%

AW_A = -80,000(A/P,19.84%,5) – 70,000 + 10,000(A/F,19.84%,5)
 = $-95,308   (PMT function)

AW_B = -125,000(A/P,19.84%,5) – 50,000 + 20,000(A/F,19.84%,5)
 = $-88,954   (PMT function)

Select Machine B.

FE Review Solutions

14.33 Answer is (a)

14.34 Answer is (d)

14.35 Answer is (b)

14.36 Answer is (c)

14.37 Answer is (b)

Extended Exercise Solution

1. Find overall i* = 5.90%.

2. i_f = 11.28%.
   F = 25,000(F/P,11.28%,3) − 1475(F/A,11.28%,3)

3. F = 25,000(F/P,4%,3)

4. Subtract the future value of each payment from the bond face value 3 years from now. Both amounts take purchasing power into account.

   F = 25,000(F/P,4%,3) − 1475[(1.04)^2 + (1.04) + 1] = $23,517

   In Excel, this can be written as:

   FV(4%,3,1475,−25000) = $23,517

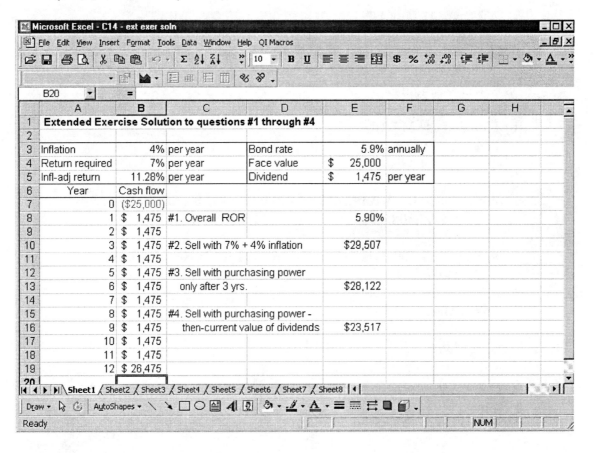

5. Use Solver to find the purchase price (B7) at 11.28% (E8).

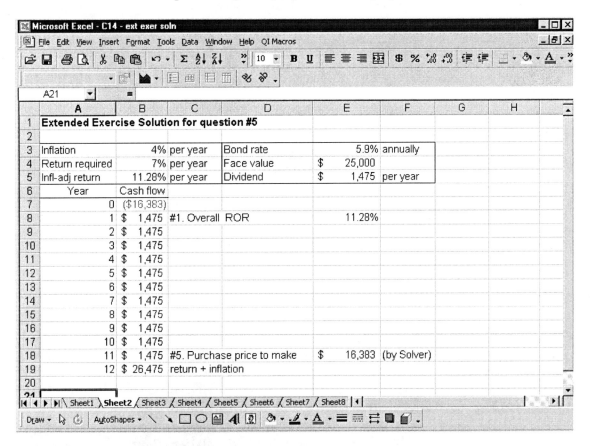

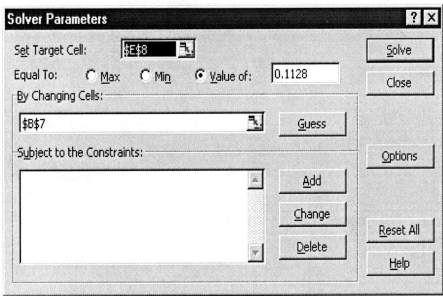

Chapter 14

Chapter 15 – Cost Estimation and Indirect Cost Allocation
Solutions to end of chapter exercises

Problems

15.1 (a) Equipment cost, delivery charges, installation cost, insurance, and training.
 (b) Labor, materials, maintenance, power.

15.2 The main differences between bottom-up and top-down approaches to cost estimating are in the input and output variables. The bottom-up approach uses price as output and cost estimates as inputs. The top-down approach is just the opposite.

15.3 The unit cost method establishes unit cost factors and uses them to estimate costs by multiplying the factors by the number of units. Examples of unit cost factors are cost/mile, $/bed, cost/ft^2, etc.

15.4 VP-Facilities: $150(130,000) = $19.5 million

 VP-Finance:

Type	Area	Unit cost	Estimated cost
Classroom	52,000	$110	$5.7200 million
Lab	45,500	165	7.5075 million
Office	32,500	90	2.9250 million
Furnishings	130,000	25	3.2500 million
			19.4025 million

Total estimate is $19,402,500, which is very close to the $19.5 million estimate.

15.5 Cost = $\dfrac{1200}{915.1}(25,000)$
 = $32,783

15.6 $30,000 = \dfrac{x}{915.1}(20,000)$
 x = 1372.7

15.7 (a) First find the percentage increase (p%) between 1985 and 1995.
 5523.13 = 4207.84 (F/P,p,10)
 1.3126 = $(1+p)^{10}$
 p% increase = 2.76%/year

 Predicted index value in 1999 = 5523.13(F/P,2.76%,4)
  = 5523.13$(1+0.0276)^4$
  = 6158.60

(b) Difference = 6158.60 − 6059.47
= 99.13 (too high)

15.8 $160,000 = \dfrac{1315}{620}(x)$
x = $75,437

15.9   Cost in 1999 = $325,000 \dfrac{(6059.47)}{(4770.03)}$
= $412,854

15.10 Divide 1999 value by the 1990 value and multiply by 100:
1999 value = (390.6/357.6)(100)
= 109.2

15.11 390.6 = 325.3(F/P,p%,14)
$1.2007 = (1+p\%)^{14}$
p% increase = 1.32 % per year

15.12 Index in 2005 = 1068.3(F/P,4%,6)
$= 1068.3(1+0.04)^6$
= 1351.7

15.13 (a) Cost = $60,000 (1+0.02)^3 (1+0.05)^7$
= $89,594

(b) $89,594 = 60,000($I_{10}$/1203)
I_{10} = 1796.36

15.14 The cost index bases the estimate on cost differences over time for a *specified value* of variables, while a CER estimates costs between *different values* of design variables.

15.15 $C_2 = 2,300 (30/15)^{0.69}$
= $3,711

15.16 $C_2 = 400,000 (20/1)^{0.63}$
= $2,640,638

15.17 $3,000,000 = 550,000 (100,000/6000)^x$
$5.4545 = (16.6667)^x$
log 5.4545 = x log 16.667
0.7367 = 1.2218 x
x = 0.60

15.18 $300{,}000 = 100{,}000(60{,}000/15{,}000)^x$
                $3 = 4^x$
          $\log 3 = 0.4771 = x \log 4 = 0.6021$
               $x = 0.79$

15.19 $250 = 55(600/Q_1)^{0.67}$
       $4.5454 = (600/Q_1)^{0.67}$
         $Q_1 = 63$ MW

15.20 $1.5 \text{ million} = 0.2 \text{ million}(Q_2/1)^{0.80}$
        $Q_2 = 12.4$ MGD

15.21   Use Equation [15.3]
       $C_2 = 50{,}000\,(60/30)^{0.24}\,(1092.0/915.1)$
         $= \$70{,}465$

15.22 C_2 in 1985 $= 160{,}000\,(1000/200)^{0.35}$
                $= \$281{,}034$

       C_2 in 1999 $= 281{,}034\,(1068.3/789.6)$
                $= \$380{,}229$

15.23   Equipment cost index is $(1092.0/1027.5)$. Cost-capacity exponent is 0.90.

       Let $C_1 =$ cost of 150 hp system in 1995.

       C_2 in 1995 $= \$220{,}000 = C_1\,(300/150)^{0.90}$
                         $C_1 = \$117{,}895$

       Update C_1 with cost index.

       $C_{2000} = C_1\,(1092.0/1027.5)$
              $= 117{,}895\,(1.0628)$
              $= \$125{,}299$

15.24 $C_T = 2.45\,(16)$
         $= \$39.2$ million

15.25   (a) $h = 1 + 1.52 + 0.31$
            $= 2.83$

         $C_T = 2.83\,(1{,}600{,}000)$
             $= 4{,}528{,}000$

(b) $h = 1 + 1.52 = 2.52$
$C_T = [1,600,000(2.52)](1.31)$
$= \$5,281,920$

15.26 (a) $h = 1 + 0.30 + 0.30 = 1.60$

Let x be the indirect cost factor.

$C_T = 450,000 = [250,000 (1.60)] (1 + x)$
$(1+ x) = 450,000/[250,000 (1.60)]$
$= 1.125$
$x = 0.125$

The indirect cost factor used is much lower than 0.40.

(b) $C_T = 250,000[1.60](1.40)$
$= \$560,000$

15.27 Indirect cost rate for $101 = \dfrac{25,000}{600} = \$41.67$ per hour

Indirect cost rate for $102 = \dfrac{50,000}{200} = \$250$ per hour

Indirect cost rate for $103 = \dfrac{25,000}{800} = \$93.75$ per hour

Indirect cost rate for $104 = \dfrac{100,000}{1,600} = \$62.50$ per hour

15.28 (a) From Eq. [15.7]:

Basis level = $\dfrac{\text{Indirect costs allocated}}{\text{Indirect cost rate}}$

Month	Basis Level	Basis
February	2,800/1.40 = 2,000	Space
March	3,400/1.33 = 2,556	Direct labor costs
April	3,500/1.37 = 2,555	Direct labor costs
May	3,600/1.03 = 3,495	Space
June	6,000/0.92 = 6,522	Material costs

(b) The only way the rate could decrease is because of the switching of the allocation basis from month to month. If a single allocation basis had been used throughout, the rate would have had to increase for each basis. For example, if space had been used for each month based on the departmental space, the monthly rate would have been:

February: $\dfrac{2,800}{2,000} = \$1.40$ per ft^2

March: $\dfrac{3,400}{2,000} = \$1.70$ per ft^2

April: $\dfrac{3,500}{3,500} = \$1.00$ per ft^2

May: $\dfrac{3,600}{3,500} = \$1.03$ per ft^2

June: $\dfrac{6,000}{3,500} = \$1.71$ per ft^2

15.29 (a) <u>Space</u>: Use Eq. [15.7] for the rate, then allocate the $34,000.
Total space in 3 depts. = 38,000 ft^2
Rate = 34,000/38,000 = $0.89 per ft^2

(b) <u>Direct labor hours:</u>
Total hours = 2,080
Rate = 34,000/2,080 = $16.35 per hour

(c) <u>Direct labor cost:</u>
Total costs = $147,390
Rate = 34,000/147,390 = $0.23 per $

15.30 <u>Housing:</u> DLH is basis; rate is $16.35
Actual charge = 16.35(480) = $7,848

<u>Subassemblies:</u> DLH is basis; rate is $16.35
Actual charge = 16.35(1,000) = $16,350

<u>Final assembly:</u> DLC is basis; rate is $0.23
Actual charge = 0.23 (12,460) = $2,866

15.31 (a) Actual charge = (rate)(actual machine hours) where the rate value is from 15.27.

Cost center	Rate	Actual hours	Actual charge	Allocation	Variance
101	$41.67	700	$29,169	$25,000	$4,169 under
102	250.00	450	112,500	50,000	62,500 under
103	93.75	650	60,938	75,000	14,062 over
104	62.50	1,400	87,500	100,000	12,500 over
			$290,107		

(b) Total variance = allocation – actual charges
= $250,000 – 290,107
= $–40,107 (under allocation)

15.32 (a) Indirect cost charge = (allocation rate) (basis level)

    Department 1:   2.50(5,000) = $ 12,500
    Department 2:   0.95(25,000) = $ 23,750
    Department 3:   1.25(44,100) = $ 55,125
    Department 4:   5.75(84,000) = $483,000
    Department 5:   3.45(54,700) = $188,715
    Department 6:   0.75(29,000) = $ 21,750
        Total actual charges = $784,840

(b) Variance = allocation – actual charges
        = 800,000 – 784,840
        = $15,160 (over allocation)

15.33 (a) Alternatives are Make and Buy. Determine the total monthly costs, TC.

TC_{make} = –DLC – materials cost – indirect costs for Housing
        – indirect costs from Testing and Engineering
    = –31,680 – 41,000 – 20,000 – 3500
    = $–96,180 per month

TC_{buy} = $–87,500 per month

Buy the components.

(b) Three alternatives are Make/old, Buy, and Make/new, meaning with new equipment.

$TC_{make/old}$ = $–96,180 per month

TC_{buy} = $–87,500 per month

$TC_{make/new}$ = –AW of equipment – DLC – materials cost
        – total indirect costs for Housing and redistribution
        from Testing and Engineering

The new indirect costs and direct labor hours for all departments are:

Department	Indirect cost	Direct labor hours
Housing	$20,000	200
Subassemblies	45,000	1,000
Final assembly	10,000	600
Testing	13,000	---
Engineering	16,000	---
Total		1,800

Chapter 15

Redistribution rate for Testing and Engineering indirect costs is based on direct labor hours:

$$\text{Redistribution rate} = \frac{\text{Testing + Engineering indirect costs}}{\text{Total direct labor hours}}$$

$$= \frac{13{,}000 + 16{,}000}{1{,}800} = \$16.11 \text{ per hour}$$

The Housing indirect cost = 200(16.11) + 20,000 = $23,222

AW of new equipment = 375,000(A/P,1%,60) + 5000
= $13,340 per month

$TC_{make/new}$ = –13,340 – 20,000 – 41,000 – 23,222
= $–97,562

Select the buy alternative.

15.34 (a) Charge = (rate)(DLH) = 4.762 (DLH)

Plant A: 4.762 (200,000) = $952,400
Plant B: $476,200
Plant C: $8,571,600

(b) Total capacity = 125,000 + 62,500 + 1,125,000 = 1,312,500

$$\text{Rate} = \frac{\$10 \text{ million}}{1.3125 \text{ million units}} = \$7.619 \text{ per unit}$$

Plant A: 7.619 (125,000) = $952,375
Plant B: $476,188
Plant C: $8,571,375

These are the same as the DLH basis.

(c) | Plant | Actual Capacity |
|---|---|
| A | 100,000/125,000 = 0.80 |
| B | 60,000/62,500 = 0.96 |
| C | 900,000/1,125,000 = 0.80 |

Plant A: 7.619 (125,000)/0.80 = $1,190,470
Plant B: 7.619(62,500)/0.96 = $496,029
Plant C: 7.619(1,125,000)/0.80 = $10,714,219

Total allocated is $12,400,718

The first methods always allocate the exact amount of the indirect cost budget. They are based on plant parameters, not performance. The numbers in part (c) will be more (ratio > 1) or less (ratio < 1) than the allocations in (a) and (b).

15.35 As the DL hours component decreases, the denominator in Eq. [15.7], basis level, will decrease. Thus, the rate for a department using automation to replace direct labor hours will increase in the computation:

$$\text{Rate} = \frac{\text{Indirect costs}}{\text{Basis level}}$$

The increased use of indirect labor for automation requires that these costs be tracked directly when possible and the remainder allocated with bases other than DLH.

15.36 (a) Rate $= \dfrac{\$1 \text{ million}}{16,500 \text{ guests}}$
= $60.61 per guest
Charge = # guests X rate

Site	A	B	C	D
Guests	3,500	4,000	8,000	1,000
Charge	$212,135	242,440	484,880	60,610

(b) Guest-nights = guests X length of stay

Total guest-nights = 35,250
Rate $= \dfrac{\$1 \text{ million}}{35,250}$
= $28.37 per guest-night

Site	A	B	C	D
Guest-night	10,500	10,000	10,000	4,750
Charge	$297,885	283,700	283,700	134,757

(c) The actual indirect cost charge to sites C and D are significantly different by the 2 methods. Another basis could be guest-dollars, that is, total amount of money a guest (or group) spends, if this could be tracked.

15.37 Rates are determined first.

$$\text{DLH rate} = \frac{\$400,000}{51,300} = \$7.80 \text{ per hour}$$

$$\text{Old cycle time rate} = \frac{\$400,000}{97.3} = \$4,111 \text{ per second}$$

$$\text{New cycle time rate} = \frac{\$400,000}{45.7} = \$8,752.74 \text{ per second}$$

Actual charges = (rate)(basis level)

Line	10	11	12
DLH basis	$156,000	99,060	145,080
Old cycle time	53,443	229,394	117,164
New cycle time	34,136	148,797	217,068

The actual charge patterns are significantly different for all 3 bases.

15.38 (a) Workforce basis rate = $200,200/1,400
= $143 per employee

CA: 143(900) = $128,700
AZ: 143(500) = $ 71,500

(b) Accident basis rate = $200,200/560
= $357.50 per accident

CA: 357.50(425) = $151,938
AZ: 357.50(135) = $ 48,262

This basis lowers the Arizona charge since it has fewer accidents per employee relative to California site.

CA: 425/900 = 0.472
AZ: 135/500 = 0.270

(c) ABC: 80% of $200,200 is $160,160

Generation-area accident basis:

Rate: $160,160/530 = $302.19 per accident

CA: 302.19(405) = $122,387
AZ: 302.19(125) = $ 37,774

Classic: 20% of $200,200 is $40,040

Employee rate = $40,040/900 = $44.49 per employee

CA: 44.49(600) = $26,693
AZ: 44.49(300) = $13,346

Total actual charges:

CA: 122,387 + 26,693 = $149,080
AZ: 37,774 + 13,346 = $ 51,120

Comparison for (a), (b) and (c):

Basis	Employees	Accidents	20% - 80% Split
CA	$128,700	$151,938	$149,080
AZ	$ 71,500	$ 48,262	$ 51,120

The difference is not great for the accident basis compared to the split-basis approach.

FE Review Solutions

15.39 $C_2 = 400,000(6059.47/4770.03)$

= $508,128

Answer is (b)

15.40 $29,860 = 25,000(I_2/789.6)$

$I_2 = 943.1$

Answer is (d)

15.41 $C_2 = 2100\,(200/50)^{0.76}$
= $6023

Answer is (c)

15.42 $Cost_{1999} = 15,000\,(1068.3/915.1)\,(2)^{0.65}$
= $27,478

Answer is (c)

Case Study #1 Solution

1. An increase in the chemical cost moves the optimum dosage to the left, or decreases the optimum dosage in Figure 15-3. For example, at a cost of $0.25 per kilogram, the optimum dosage is about 4.7 mg/L (by trial and error using spreadsheet and total cost equation of $C_T = -0.0024F^3 + 0.0749F^2 - 0.548F + 3.791$).

2. An increase in backwash water cost raises the backwash water cost line and moves the optimum dosage to the right in Figure 15-3. For example, doubling the cost of water from $0.0608/m^3 to $0.1216/m^3$ moves the optimum dosage to 7.2 mg/L (by trial and error).

3. The chemical cost at 10 mg/L is $1.83/1000 m^3 of water produced

4. The backwash water cost at 14 mg/L is $0.71/1000 m^3 of water produced by using 14 mg/L in Eq. [15.10].

5. For $C_C = 0.21$ in Eq. [15.11], C_T in Eq. [15.12] is:
$C_T = -0.0024F^3 + 0.0749F^2 - 0.588F + 3.791$.
At 6 mg/L, total cost is: $C_T = \$2.44$.

6. The minimum dosage would be 8 mg/L at a chemical cost of $0.06/kg. Determined by trial and error using $C_T = -0.0024F^3 + 0.0749F^2 - 0.738F + 3.791$.

Case Study #2 Solution

1. DLH basis

   Standard: rate = $\dfrac{\$1.67 \text{ million}}{187{,}500 \text{ hrs}}$ = $8.91/DLH

   Premium: rate = $\dfrac{\$3.33 \text{ million}}{125{,}000 \text{ hrs}}$ = $26.64/DLH

Model	IDC rate	DL hours	IDC allocation	Direct material	Direct Labor	Total cost	Price, ~1.10 x cost
Std	$8.91	0.25	$2.23/un	2.50/m	$5/un	$9.73	10.75
Prm	26.64	0.50	13.32	3.75	10	27.07	29.75

2.

Cost pool	Cost driver	Volume of driver	Total cost/year	Cost per activity
Quality	inspections	20,000	$800,000	$40/inspection
Purchasing	orders	40,000	1,200,000	30/order
Scheduling	orders	1,000	800,000	800/order
Prod. Set-ups	set-ups	5,000	1,000,000	200/set-up
Machine Ops.	hours	10,000	1,200,000	120/hour

ABC allocation

Driver	Standard Activity	Standard IDC allocation	Premium Activity	Premium IDC allocation
Quality	8,000@$40	$320,000	12,000@$40	$480,000
Purchasing	30,000@30	900,000	10,000@30	300,000
Scheduling	400@800	320,000	600@800	480,000
Prod. Set-ups	1,500@200	300,000	3,500@200	700,000
Machine Ops.	7,000@120	840,000	3,000@120	360,000
Total		$2,680,000		$2,320,000
Sales volume		750,000		250,000
IDC/unit		$3.57		$9.28

Model	Direct material	Direct labor	IDC allocation	Total cost
Standard	2.50	5.00	3.57	$11.07
Premium	3.75	10.00	9.28	$23.03

3. Traditional

Model	Profit/unit	Volume	Profit
Standard	10.75 – 9.73 = $1.02	750,000	$765,000
Premium	29.75 – 27.07 = $2.68	250,000	670,000
Profit			$1,435,000

ABC

Model	Profit/unit	Volume	Profit
Standard	10.75 – 11.07 = $–0.32	750,000	$ –240,000
Premium	29.75 – 23.03 = $6.72	250,000	1,680,000
Profit			$1,440,000

4. Price at Cost + 10%

Model	Cost	Price	Profit/unit	Volume	Profit
Standard	$11.07	$12.18	$1.11	750,000	$832,500
Premium	23.03	25.33	2.30	250,000	575,000
Profit					$1,407,000

Profit goes down ~$33,000

5. a) They were right on IDC allocation under ABC, but they were wrong on traditional where the cost is ~ 1/3 and IDC is ~1/6.

| | Allocation | |
Model	Traditional	ABC
Standard	$2.23/unit	$3.57/unit
Premium	13.32	9.28

b) <u>Cost versus Profit comment</u> – Wrong if old prices are retained.
Under ABC standard model loses $0.32/unit. Price for standard should go up.
Price for standard should go up. Premium makes good profit at current price under ABC ($7.72/unit).

c) <u>Premium require more activities and operations</u>
Wrong : <u>Premium is lower</u> in cost drivers of purchase orders and machine operations hours, but is higher on set ups and inspections. However, number of set-ups is low (5000 total) and (quality) inspections have a low cost at $40/inspection.
Overall – Not a correct impression when costs are examined.

Chapter 16 – Depreciation Methods
Solutions to end of chapter exercises

Problems

16.1 Book depreciation is used for internal records to reflect current capital investment in the asset. Tax depreciation is used to determine the annual tax-deductible amount. They are not necessarily the same.

16.2 MACRS has set n values for depreciation by property class. These are commonly different – usually shorter – than the actual, anticipated useful life of an asset.

16.3 (a) Depreciation is the decrease in the value of property over the time the property is being used. Events that can cause property to depreciate include wear and tear, age, deterioration, and obsolescence.

(b) General Depreciation System (GDS) and Alternative Depreciation System (ADS). The recovery period and method of depreciation are the primary differences.

(c) Conventions are mid-month, mid-quarter, and half-year.

(d) The following cannot be MACRS depreciated: intangible property, motion picture film and video tape, sound recordings, certain real and personal property placed into service before 1987.

16.4 (a) B = \$270,000 + 30,000 = \$300,000
n = 10 years
S = 0.1(270,000) = \$27,000

(b) Remaining life = 5 years
Market value = \$25,000
Book Value = \$300,000 – 0.6(300,000)
= \$105,000

16.5

year	Book Value	Part (a) Annual Depreciation	Part (b) Depreciation Rate
0	\$100,000	0	-----
1	90,000	\$10,000	10 %
2	81,000	9000	9
3	72,900	8100	8.1
4	65,610	7290	7.3
5	59,049	6561	6.56

(c) Book value = \$59,049 and market value = \$24,000.

16.6 B = $350,000; S = 0.1(325,000) = $32,500
D_t=(350,000-32,500)/7 = $45,357.14 per year t
BV_5 = 350,000 – 5(45,357.14) = $123,214.29

16.7 (a) $D_t = \dfrac{12,000 - 2000}{8} = \$1250$

(b) BV_3 = 12,000 – 3(1250) = $8250

(c) d = 1/n = 1/8 = 0.125

16.8 (a) B = $50,000, n = 4, S = 0, d = 0.25

Year$_t$	D_t	Accumulated D_t	BV_t
0	----------	----------	$50,000
1	$12,500	$12,500	37,500
2	12,500	25,000	25,000
3	12,500	37,500	12,500
4	12,500	50,000	0

(b) S = $16,000, d = 0.25, B - S = $34,000

Year	D_t	Accumulated D_t	BV_t
0	-------	-------	$50,000
1	$8,500	$8,500	41,500
2	8,500	17,000	33,000
3	8,500	25,500	24,500
4	8,500	34,000	16,000

Plot year versus D_t, accumulated D_t and BV_t on one graph for each salvage value.

(c) Spreadsheets for S = 0 and S = $16,000 provide the same answers as above.

16.9 = 100,000 – 3 * SLN(100000,20000,8).
Answer is $70,000.

16.10 d is amount of BV removed each year.

d_{max} is maximum legal rate of depreciation for each year; 2/n for DDB.

d_t is actual depreciation rate charged using a particular depreciation model; for DB model it is $d(1-d)^{t-1}$.

16.11

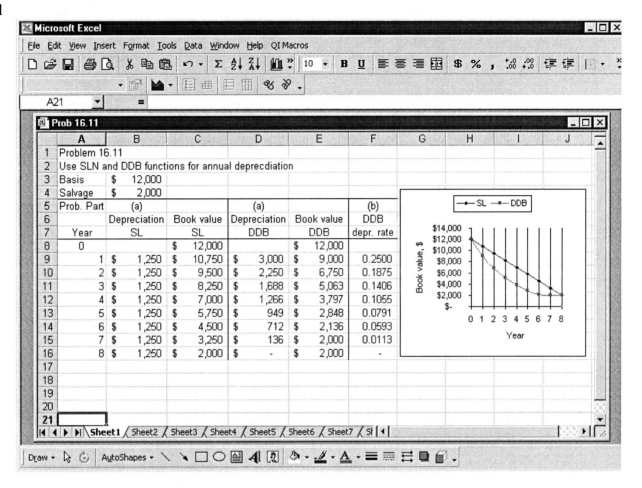

16.12 B = $600,000; n = 30; S = 0

(a) Straight line depreciation:

$$D_t = \frac{600,000}{30} = \$20,000 \qquad t = 1,2,\ldots,30$$

(b) Double declining balance method: d = 2/n = 2/30 = 0.06667

$$D_4 = 0.06667(600,000)(1-0.06667)^{4-1} = \$32,523$$

$$D_{10} = 0.06667(600,000)(1-0.06667)^{10-1} = \$21,498$$

$$D_{25} = 0.06667(600,000)(1-0.06667)^{25-1} = \$7637$$

The annual depreciation values are significantly different:

Year	SL	DDB
4	$20,000	$32,523
10	20,000	21,498
25	20,000	7,637

(c) $D_{30} = 600,000(1-0.06667)^{30} = \$75,720$

16.13 SL: $d_t = 0.20$ of B = $25,000
$BV_t = 25,000 - t(5,000)$

Fixed rate: DB with $d = 0.25$
$BV_t = 25,000(0.75)^t$

DDB: $d = 2/5 = 0.40$
$BV_t = 25,000(0.60)^t$

		Declining balance methods	
Year, t	SL	125% SL	200% SL
d	0.20	0.25	0.40
0	$25,000	$25,000	$25,000
1	20,000	18,750	15,000
2	15,000	14,062	9,000
3	10,000	10,547	5,400
4	5,000	7,910	3,240
5	0	5,933	1,944

16.14 For DDB, use $d = 2/18 = 0.11111$

$D_2 = 0.11111(182,000)(1 - 0.11111)^{2-1} = \$17,975$

$D_{18} = 0.11111(182,000)(1 - 0.11111)^{18-1} = \$2730$

Compare BV_{17} with S = $50,000. By Eq. [16.8]

$BV_{17} = 182,000(1 - 0.11111)^{17} = \$24,575$

It is not okay to use $D_{18} = \$2730$ because the BV has already reached the estimated S of $50,000. In fact, $BV_{10} = \$56,047$ so $D_{11} = \$6227$ cannot be fully applied.

For DB, calculate d via Eq. [16.11].

$$d = 1 - (50{,}000/182{,}000)^{1/18} = 0.06926$$

$$D_2 = 0.06926(182{,}000)(0.93074)^1 = \$11{,}732$$

$$D_{18} = 0.06926(182{,}000)(1-0.06926)^{18-1} = \$3721$$

16.15 The implied d is 0.06926. The factor for the DDB function is
factor = implied DB rate / SL rate
= 0.06926 / (1/18)
= 1.24668

The DDB function is DDB(182000,50000,18,18,1.24668)

$$D_{18} = 0.06926(182{,}000)(0.93074)^{17} = \$3721$$

The D_{18} value must be acceptable since d was calculated using estimated values.

16.16 (a) $d = \dfrac{1.5}{12} = 0.125$

$$D_1 = 0.125(175{,}000)(0.875)^{1-1} = \$21{,}875$$

$$BV_1 = 175{,}000(0.875)^1 = \$153{,}125$$

$$D_{12} = 0.125(175{,}000)(0.875)^{12-1} = \$5{,}035$$

$$BV_{12} = 175{,}000(0.875)^{12} = \$35{,}248$$

(b) The 150% DB salvage value of \$35,248 is larger than S = \$32,000.

16.17 Personal property: manufacturing equipment, tractor, company car
Real property: rental house, warehouse building (not land of any kind)

16.18

Year	SL*		MACRS		
t	D_t	BV_t	d_t	D_t	BV_t
0	----	$30,000	-----	----	$30,000
1	$4,000	26,000	0.1429	$4,287	25,713
2	4,000	22,000	0.2449	7,347	18,366
3	4,000	18,000	0.1749	5,247	13,119
4	4,000	14,000	0.1249	3,747	9,372
5	4,000	10,000	0.0893	2,679	6,693
6	4,000	6,000	0.0892	2,676	4,017
7	4,000	2,000	0.0893	2,679	1,338
8	0	2,000	0.0446	1,338	0

* $D_t = (30,000 - 2000)/7 = \$4000$

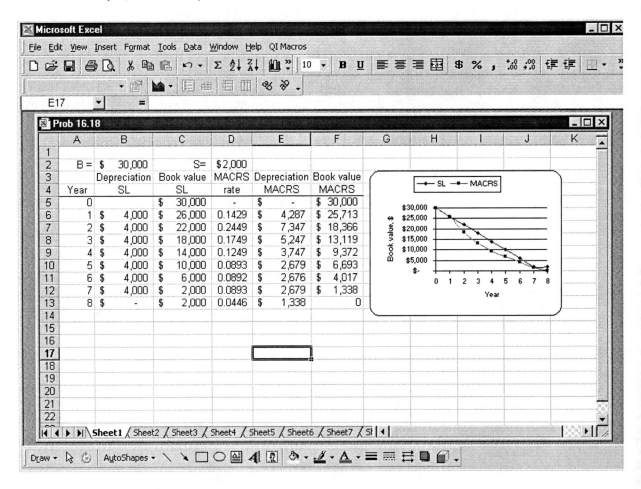

16.19 For MACRS use Table 16.2 rates for n = 5. For DDB, with d = 0.2, stop depreciating at S = $4,000.

	(a) MACRS			(b) DDB	
Year t	d_t	D_t	BV_t	D_t	BV_t
0	--	-----	$20,000	-----	$20,000
1	0.20	$4,000	16,000	$4,000	16,000
2	0.32	6,400	9,600	3,200	12,800
3	0.192	3,840	5,760	2,560	10,240
4	0.1152	2,304	3,456	2,048	8,192
5	0.1152	2,304	1,152	1,638	6,554
6	0.0576	1,152	0	1,311	5,243
7	-------		0	1,048	4,195
8	-------		0	195*	4,000
9	-------		0	0	4,000
10	-------		0	0	4,000

*D_t = 0.2(4195) = $839 not allowed
MACRS depreciates to S = 0 while DDB stops at $4,000.
Plot the two BV_t columns on an x–y scatter chart.

16.20 For classical SL, n = 5 and
D_t = 450,000/5 = $90,000
BV_3 = 450,000 − 270,000 = $180,000

For MACRS, after 3 years for n = 5 sum the d_t values in Table 16.2.
ΣD_t = 450,000(0.712) = $320,400

BV_3 = $450,000 − 320,000 = $129,600

The difference is $50,400, which has not been removed by the classical SL method.

16.21 Use n = 39 with d_t = 1/39 = 0.02564 in all years except 1 and 40 as specified by MACRS.

Year, t	d_t	D_t
1	0.01391	$2,086.50
2-39	0.02564	3,846.00
40	0.01177	1,765.50

16.22 (a) For MACRS, use n = 5 and the Table 16.2 rates with B = $100,000.
For SL, use n = 10 with d_t = 0.05 in years 1 and 11.

Year	MACRS			Modified SL		
t	d_t	D_t	BV_t	d_t	D_t	BV_t
0	----	----	$100,000	-----	------	$100,000
1	0.2000	$20,000	80,000	0.05	$ 5,000	95,000
2	0.3200	32,000	48,000	0.10	10,000	85,000
3	0.1920	19,200	28,800	0.10	10,000	75,000
4	0.1152	11,520	17,280	0.10	10,000	65,000
5	0.1152	11,520	5760	0.10	10,000	55,000
6	0.0576	5760	0	0.10	10,000	45,000
7	--------	------	0	0.10	10,000	35,000
8	--------	------	0	0.10	10,000	25,000
9	--------	------	0	0.10	10,000	15,000
10	--------	------	0	0.10	10,000	5000
11	--------	------	0	0.05	5000	0

Plot the two BV_t column values on one graph.

(b) MACRS: sum d_t values for 3 years: 0.20 + 0.32 + 0.192 = 0.712 (71.2%)
SL: sum the d_t values for 3 years: 0.05 + 0.1 + 0.1 = 0.25 (25%)
SL depreciates much slower early in the recovery period.

16.23

Year	MARCS		Modified SL	
t	d_t	D_t	d_t	D_t
1	0.20	$4000	0.0555	$1110
2	0.32	6400	0.1111	2222
3	0.192	3840	0.1111	2222
4	0.1152	2304	0.1111	2222
5	0.1152	2304	0.1111	2222
6	0.0576	1152	0.1111	2222
7	---------	-------	0.1111	2222
8	---------	-------	0.1111	2222
9	---------	-------	0.1111	2222
10	---------	-------	0.0555	1110

The SL rate is 1/9 = 0.1111 except in years 1 and 10 when it is 1/18 = 0.0555.

16.24

t	SL	MACRS	SL Alternative
1	33.3	33.33	16.7
2	33.3	44.45	33.3
3	33.3	14.81	33.3
4	0	7.41	16.7

d_t values (%)

Plot the 3 curves.

16.25 There is more depreciation, which is a tax deduction, so more revenue is retained as net profit after taxes.

16.26 (a) 15 years
(b) 3 years
(c) 7 years
(d) 5 years

16.27 Percentage depletion for copper is 15% of gross income, not to exceed 50% of taxable income.

Year, t	Gross* income	% Depl @ 15%	50% of TI	Allowed depletion
1	$3,200,000	$480,000	$750,000	$480,000
2	7,020,000	1,053,000	1,000,000	1,000,000
3	2,990,000	448,500	500,000	448,500

*(Tons)($/pound)(2000 pounds/ton)

16.28 (a) p_t = $3.2/2 million = $1.6 per ton
Percentage depletion is 5% of gross income each year

Year, t	Tonnage for cost depletion	Gross income for percentage depletion
1	50,000	$ 750,000
2	42,000	630,000
3	58,000	1,160,000
4	60,000	1,200,000
5	65,000	1,560,000

Year, t	$Depl, $1.6 x tonnage per year	%Depl, 5% of GI	Selected
1	$80,000	$37,500	$Depl
2	67,200	31,500	$Depl
3	92,800	58,000	$Depl
4	96,000	60,000	$Depl
5	104,000	78,000	$Depl

(b) Total depletion is $440,000
% written off = 440,000/3.2 million = 13.75%

(c) The undepleted investment after 3 years:
3.2 million – (80,000 + 67,200 + 92,800) = $2,960 million

New cost factor is:

p_t' = $2.960 million/2.5 million tons
  = $1.184 per ton

Cost depletion for years 4 and 5:
year 4: 60,000(1.184) = $71,040
year 5: 65,000(1.184) = $76,960

Percentage depletion amounts are the same.
Conclusion: Select $Depl for year 4 and %Depl in year 5.

t	1	2	3	4	5
Depletion	$80,000	$67,200	$92,800	$71,040	$78,000

% written off = $389,040/3.2 million = 12.16%

FE Review Solutions

16.29 $D = \dfrac{20,000 - 2000}{5}$ = $3600 per year

Answer is (a)

16.30 From table, depreciation factor is 14.40%.

D = 30,000(0.144) = $4320

Answer is (b)

16.31 $D = \dfrac{50{,}000 - 10{,}000}{5} = \$8000$ per year

$BV_3 = 50{,}000 - 3(8{,}000) = \$26{,}000$

Answer is (d)

16.32 The MACRS depreciation rates are 0.2 and 0.32.

$D_1 = 50{,}000(0.20) = \$10{,}000$

$D_2 = 50{,}000(0.32) = \$16{,}000$

$BV_2 = 50{,}000 - 10{,}000 - 16{,}000 = \$24{,}000$

Answer is (c)

16.33 By the straight line method, book value at end of asset's life MUST equal salvage value ($10,000 in this case).

Answer is (c)

16.34 Total depreciation = first cost − BV after 3 years
= 50,000 − 21,850 = $28,150

Answer is (d)

Chapter 16 Appendix Solutions

16A.1 The depreciation rate is from Eq. [16A.4] using SUM = 36.

t	d_t	D_t, euro	BV_t, euro
1	8/36	2,222.22	9777.78
2	7/36	1,944.44	7833.33
3	6/36	1,666.67	6166.67
4	5/36	1,388.89	4777.78
5	4/36	1,111.11	3666.67
6	3/36	833.33	2833.33
7	2/36	555.56	2277.78
8	1/36	277.78	2000.00

$BV_1 = 12{,}000 - \left[\dfrac{1(8 - 0.5 + 0.5)}{36}\right](12{,}000 - 2000) = 9777.78$ euro

$$BV_2 = 12{,}000 - \left[\frac{2(8-1+0.5)}{36}\right](10{,}000) = 7833.33 \text{ euro}$$

16A.2 Use B = $150,000; n = 10; S = $15,000 and SUM = 55.

$$D_2 = \frac{10-2+1}{55}(150{,}000 - 15{,}000) = \$22{,}091$$

$$BV_2 = 150{,}000 - \left[\frac{2(10-1+0.5)}{55}\right](150{,}000 - 15{,}000) = \$103{,}364$$

$$D_7 = \frac{10-7+1}{55}(150{,}000 - 15{,}000) = \$9818$$

$$BV_7 = 150{,}000 - \left[\frac{7(10-3.5+0.5)}{55}\right](150{,}000 - 15{,}000) = \$29{,}727$$

16A.3 B = $12,000; n = 6 and S = 0.15(12,000) = $1,800

(a) Use Eq. [16A.3] and S = 21.
$$BV_3 = 12{,}000 - \left[\frac{3(6-1.5+0.5)}{21}\right](12{,}000 - 1800) = \$4714$$

(b) By Eq. [16A.4] and t = 4:

$$d_4 = \frac{6-4+1}{21} = 3/21 = 1/7$$

$$\begin{aligned}D_4 &= d_4(B-S)\\ &= (3/21)(12{,}000 - 1800)\\ &= \$1457\end{aligned}$$

16A.4 B = $45,000   n = 5   S = $3000   i = 18%

Compute the D_t for each method and select the larger value to maximize PW_D.

For DDB, d = 2/5 = 0.4. By Eq. [16A.6]
$BV_5 = 45{,}000(1 - 0.4)^5 = 3499 > 3000$

Switching is advisable.

	DDB Method		Switching to SL method	Larger
t	Eq. [16A.7]	BV_t	Eq. [16A.8]	D_t
0	------------	$45,000	------------	---------
1	$18,000	27,000	$8,400	$18,000 (DDB)
2	10,800	16,200	6,000	10,800 (DDB)
3	6,480	9,720	4,400	6,480 (DDB)
4	3,888	5,832	3,360	3,888 (DDB)
5	2,333	3,499	2,832	2,832 (SL)

The switch to SL occurs in year 5 and the PW of depreciation is:

$PW_D = 18,000(P/F,18\%,1) + \ldots + 2,832(P/F,18\%,5)$
$= \$30,198$

16A.5 Develop a spreadsheet for the DDB-to-SL switch and MACRS values plus the PW_D.

From Problem 16A.4 above, switching in year 5 from DDB to SL results in:
$PW_D = \$30,198$

For MACRS, the schedule is:

t	d_t	D_t
1	0.200	$ 9000
2	0.320	14,400
3	0.192	8640
4	0.1152	5184
5	0.1152	5184
6	0.0576	2592
		$45,000 $PW_D = \$29,128$

Were switching allowed in the USA, it would give a slightly higher PW_D.

16A.6 175% DB: $d = \dfrac{1.75}{10} = 0.175$ for t = 1 to 5

$BV_t = 110,000(0.825)^t$

SL: $D_t = \dfrac{BV_5 - 10,000}{5}$

$= (42,040 - 10,000)/5 = \$6408$ for t = 6 to 10
$BV = BV_5 - t(6408)$

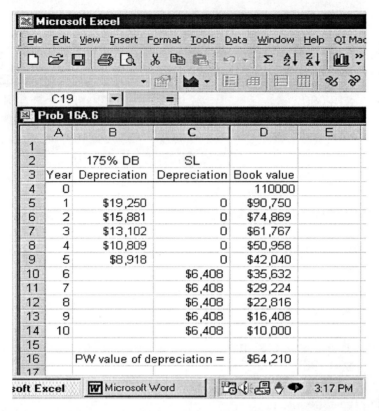
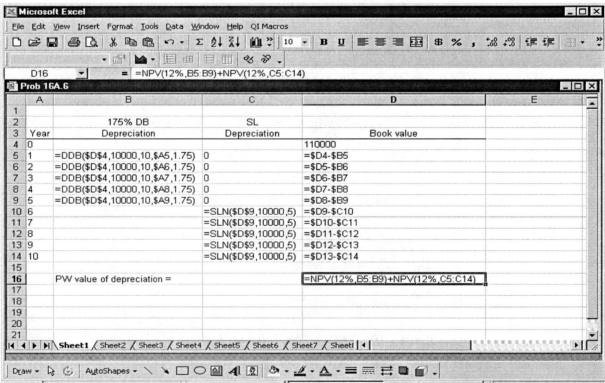

16A.7 (a) Solution gives a value $PW_D = \$6511$ for $d = 0.25$.

t	D_t for DDB	BV_t	(P/F,20%,t)	PW_D for DDB
0	------------	$12,000.00	--------------	--------------
1	$3000.00	9000.00	0.8333	$2499.90
2	2250.00	6750.00	0.6944	1562.48
3	1687.50	5062.50	0.5787	976.56
4	1265.63	3796.88	0.4828	610.41
5	949.22	2847.66	0.4019	381.49
6	711.92	2135.75	0.3349	238.44
7	533.94	1601.81	0.2791	149.02
8	400.45	1201.36	0.2326	93.14
				$6511.36

(b) Switch to SL from DDB in year 7 with $PW_D = \$6611$.

t	D_t	Method	BV_t
0	------------	--------	$12,000.00
1	$3000.00	DDB	9000.00
2	2250.00	DDB	6750.00
3	1687.50	DDB	5062.00
4	1265.63	DDB	3796.88
5	949.22	DDB	2847.66
6	711.92	DDB	2135.75
7	667.87	SL	1467.88
8	667.87	SL	800.00

(c) For MACRS with n = 7, PW_D = $6620

MACRS

t	d_t	D_t	BV_t
0	------	------	$12,000.00
1	0.1429	1714.80	10,285.20
2	0.2449	2938.80	7346.40
3	0.1749	2098.80	5247.60
4	0.1249	1498.80	3748.80
5	0.0893	1071.60	2677.20
6	0.0892	1070.40	1606.80
7	0.0893	1071.60	535.20
8	0.0446	535.20	0

All the PW_D values are very close to each other.

(d) Plot the three BV_t column values.

16A.8 (a) Use Eq. [16A.6] for DDB with d = 2/25 = 0.08.

$$BV_{25} = 155,000(1 - 0.08)^{25} = \$19,276.46 < 50,000$$

No, the switch should not be made.

(b) $155,000(1-d)^{25} > 50,000$

$$1 - d > \left[\frac{50,000}{155,000}\right]^{1/25}$$
$$1 - d > (0.3226)^{0.04} = 0.95575$$
$$d < 1 - 0.95575 = 0.04425$$

If d < 0.04425 the switch is advantageous.

16A.9 Answer depends on country in which learner/student resides.

16A.10 Verify that the rates are the following with d = 0.40:

t	1	2	3	4	5	6
d_t	0.20	0.32	0.192	0.1152	0.1152	0.0576

d_1: $d_{DB,1} = 0.5d = 0.20$

d_2:   By Eq. [16A.14] for DDB:

$d_{DB,2} = 0.4(1 - 0.2) = 0.32$ (Selected)

By Eq. [16A.15] for SL:

$d_{SL,2} = 0.8/4.5 = 0.178$

d_3:   For DDB

$d_{DB,3} = 0.4(1 - 0.2 - 0.32)$
$= 0.192$ (Selected)

For SL
$d_{SL,2} = 0.48/3.5 = 0.137$

d_4: $d_{DB,4} = 0.4(1 - 0.2 - 0.32 - 0.192)$
$= 0.1152$
$d_{SL,4} = 0.288/2.5 = 0.1152$ (Select either)

Switch to SL occurs in year 4.

d_5:   Use the SL rate n = 5.
$d_{SL,5} = 0.1728/1.5 = 0.1152$

d_6: $d_{SL,6}$ is the remainder or 1/2 the d_5 rate.

$d_{SL,6} = 1 - \sum_{t=1}^{5} d_t = 1 - (0.2 + 0.32 + 0.192 + 0.1152 + 0.1152)$

$= 0.0576$

16A.11 B = $30,000   n = 5 years   d = 0.40

Find BV_3 using d_t rates derived or rates from Table 16.2.

This answer derives the rates:

t = 1: $d_1 = 1/2(0.4) = 0.2$
$D_1 = 30,000(0.2) = \$6000$
$BV_1 = \$24,000$

t = 2: For DDB depreciation, use Eq. [16A.11]
$d = 0.4$
$D_{DB} = 0.4(24,000)$
   $= \$9600$

$BV_2 = 24,000 - 9600 = \$14,400$

For SL, if switch is better, in year 2, by Eq. [16A.12].

$D_{SL} = \dfrac{24,400}{5-2+1.5} = \$5333$

Select DDB; it is larger.

t = 3: For DDB, apply Eq. [16A.11] again.

$D_{DB} = 14,400(0.4) = \$5760$
$BV_3 = 14,400 - 5760 = \$8640$

For SL, Eq. [16A.12]

$D_S = \dfrac{14,400}{5-3+1.5} = \$4114$

Select DDB.
Conclusion: When sold for \$5000, $BV_3 = \$8640$. Therefore, there is a loss of \$3640 relative to the MACRS book value.

NOTE: If Table 16.2 rates are used:
  Cumulative depreciation in % for 3 years is:

$20 + 32 + 19.2 = 71.2\%$
$30,000(0.712) = \$21,360$

$BV_3 = 30,000 - 21,360 = \$8640$

16A.12 Determine MACRS rates for n = 7 and apply them to B = \$50,000. (S) indicates the selected method and amount.

DDB	SL
t = 1: $d = 1/7 = 0.143$	$D_{SL} = \dfrac{50,000}{7-1+1.5} = \$6667$
$D_{DB} = \$7150$ (S)	
$BV_1 = \$42,850$	

$t = 2$: $d = 2/7 = 0.286$ $D_{SL} = \dfrac{42{,}850}{7-2+1.5} = \$6592$
$D_{DB} = \$12{,}255$ (S)
$BV_2 = \$30{,}595$

$t = 3$: $d = 0.286$ $D_{SL} = \dfrac{30{,}595}{7-3+1.5} = \$5563$
$D_{DB} = \$8750$ (S)
$BV_3 = \$21{,}845$

$t = 4$: $d = 0.286$ $D_{SL} = \dfrac{21{,}845}{7-4+1.5} = \$4854$
$D_{DB} = \$6248$ (S)
$BV_4 = 15{,}597$

$t = 5$: $d = 0.286$ $D_{SL} = \dfrac{15{,}597}{7-5+1.5} = \$4456$
$D_{DB} = \$4461$ (S)
$BV_5 = \$11{,}136$

$t = 6$: $d = 0.286$ $D_{SL} = \dfrac{11{,}136}{7-6+1.5} = \$4454$ (S)
$D_{DB} = \$3185$
(Use SL hereafter) $BV_6 = \$6682$

$t = 7$: $D_{SL} = \dfrac{6682}{7-7+1.5} = \$4454$

$BV_7 = \$2228$

$t = 8$: $D_{SL} = \$2228$
$BV_8 = 0$

The depreciation amounts sum to $50,000.

t	D_t	t	D_t
1	$7150	5	$4461
2	12,255	6	4454
3	8750	7	4454
4	6248	8	2228

16A.13 (a) The SL rates with the half-year convention for n = 3 are:

t	d_t	Formula
1	0.167	1/2n
2	0.333	1/n
3	0.333	1/n
4	0.167	1/2n

(b)

t	1	2	3	4	PW_D
MACRS	$26,664	35,560	11,848	5928	$61,253
SL Alternative	$13,360	26,640	26,640	13,360	$56,915

The MACRS PW_D is larger by $4338.

Chapter 17 – After-Tax Economic Analysis
Solutions to end of chapter exercises

Problems

17.1   GI = GI
       TI = GI – E – D
       NI = (GI – E – D)(1 – T)
       GI in all 3; E and D in 2; and T in 1.

17.2   (a) Depreciation
       (b) Net profit after taxes
       (c) Taxable income
       (d) Gross income
       (e) Taxable income
       (f) Expenses
       (g) Gross income and taxable income

17.3   Graduated tax rate means the taxpayer pays at a higher rate as the taxable income increases. Marginal tax rate is the portion or percentage of each new dollar of taxable income that is paid in income taxes.

17.4   (a) <u>Company 1</u>
       TI   = Gross income - Expenses - Depreciation
            = (1,500,000 + 31,000) – 754,000 – 48,000
            = $729,000
       Taxes = 113,900 + 0.34(729,000 – 335,000)
            = $247,860

       <u>Company 2</u>
       TI   = (820,000 + 25,000) – 591,000 – 18,000
            = $236,000
       Taxes = 22,250 + 0.39(236,000 – 100,000)
            = $75,290

       (b) Co. 1:   247,860/1.5 million = 16.5%
           Co. 2:   75,290/820,000 = 9.2%

       (c) <u>Company 1</u>
           Taxes = (TI)(T_e) = 729,000(0.34) = $247,860
           % error with graduated tax = .0%

           <u>Company 2</u>
           Taxes = 236,000(0.34) = $80,240

           % error = $\frac{80,240 - 75,290}{75,290}$ (100%) = +6.6%

17.5　Taxes using graduated rates:

$$\text{Taxes on } \$300{,}000: 22{,}250 + 0.39(200{,}000) = \$100{,}250$$

(a) Average tax rate = 100,250/300,000 = 34.0%

(b) 34% from Table 17.1

(c) Taxes = 113,900 + 0.34(165,000) = $170,000
Average tax rate = 170,000/500,000 = 34.0%

(d) Marginal rate is 39% for $35,000 and 34% for $165,000. Use Eq. [17.3].
NPAT = 200,000 − 0.39(35,000) − 0.34(165,000) = $130,250

17.6　$T_e = 0.076 + (1 − 0.076)(0.34) = 0.390$
TI = 6.5 million − 4.1 million = $2.4 million
Taxes = 2,400,000(0.390) = $936,000

17.7　(a) Taxes = 13,750 + 0.34(5000) = $15,450
Average federal rate = $\dfrac{15{,}450}{80{,}000}(100\%)$
= 19.3%

(b) Effective tax rate = 0.06 + (1 − 0.06)(0.193)
= 0.2414

(c) Total taxes = 80,000(0.2414) = $19,314

(d) State: 80,000(0.06) = $4800
Federal: 80,000[0.193(1 − 0.06)] = 80,000(0.1814) = $14,514

17.8　Without system
Taxes = 150,000(0.39) = $58,500

With system
D = $8000
TI = 150,000 + 9000 − 2000 − 8000 = $149,000
Taxes = 149,000(0.39) = $58,110

Tax difference = 58,500 − 58,110 = $390 (reduction)

17.9　(a) GI = 98,000 + 7500 = $105,500
TI = 105,500 − 10,500 = $95,000

Using the rates in Table 17.2:

Taxes = 0.15(26,250) + 0.28(63,550 − 26,250) + 0.31(95,000 − 63,550)
= 0.15(26,250) + 0.28(37,300) + 0.31(31,450)
= $24,131

(b) 24,131/98,000 = 24.6%

(c) Based on TI,
24,131/95,000 = 25.4%

17.10 Single:
TI = 60,000 − 2800 − 7000 = $50,200
Taxes = 0.15(26,250) + 0.28(23,950) = $10,644
Joint:
TI = 60,000 − 5600 − 7000 = $47,400
Taxes = 0.15(43,850) + 0.28(3550) = $7572

Individuals pay more taxes since the marginal tax-rate ranges are lower than for joint filing. See Table 17.2.

17.11 NPAT = GI − E − D − taxes
CFAT = GI − E − P + S − taxes

The NPAT expression deducts depreciation outside the TI and tax computation. The CFAT expression removes the capital investment (or adds the salvage) but does not consider depreciation, since it is a noncash flow.

17.12 Depreciation is only used to find TI. Depreciation is not a true cash flow, and as such is not a direct reduction when determining either term for an alternative.

17.13 (a) CFAT = GI − E − P + S − taxes

(b) NPAT = TI − taxes

All numbers are times $10,000.

Year	GI	E	P or S	D	TI	Taxes	(a) CFAT	(b) NPAT
0	−	−	−20	−	−	−	$-20.0	
1	8	2		4	2	0.8	5.2	1.2
2	15	4		4	7	2.8	8.2	4.2
3	12	3		4	5	2.0	7.0	3.0
4	10	5	4	4	1	0.4	8.6	0.6

17.14 MACRS rates with n = 3 are from Table 16-2. All numbers are times $10,000.

	Year	GI	P or S	E	D	TI	Taxes	(a) CFAT	(b) CFAT
	0	–	-20	–	–	–	–	-20.000	-20.000
	1	8		2	6.666	-.666	-.266	6.266	6.266
	2	15		4	8.890	2.110	.844	10.156	10.156
	3	12		3	2.962	6.038	2.415	6.585	6.585
(a)	4	10	0	5	1.482	3.518	1.407	3.593	–
(b)	4	10	2	5	1.482	3.518	1.407	–	5.593

The S = $20,000 in year 4 is positive cash flow. CFAT for years 0 through 3 are the same as
for S = 0.

17.15 No capital purchase (P) or salvage (S) is involved.

$$\begin{aligned} CFBT &= CFAT + taxes \\ &= CFAT + TI(T_e) \\ &= CFAT + (GI - E - D)T_e \\ &= CFAT + (CFBT - D)T_e \end{aligned}$$

$$CFBT = [CFAT - D(T_e)]/(1 - T_e)$$

$T_e = 0.045 + 0.955(0.35) = 0.37925$

$$\begin{aligned} CFBT &= [2{,}000{,}000 - (1{,}000{,}000)(0.37925)]/(1 - 0.37925) \\ &= 1{,}620{,}750/0.62075 \\ &= \$2{,}610{,}955 \end{aligned}$$

17.16 CFBT = GI – Expenses – Investment + Salvage
       TI = CFBT – Depreciation
       Taxes = 0.4(TI)
       CFAT = CFBT – taxes
       NPAT = TI – taxes

	A	B	C	D	E	F	G	H	I	J
1		= Interest			(All values are $)					
2	40%	= Tax rate								
3		Gross		Investment			Taxable			
4		income	Expenses	or salvage		Depreciation	income			
5	Year	GI	E	P or S	CFBT	D	TI	Taxes	NPAT	CFAT
6	0			(250,000)	(250,000)				0	(250,000)
7	1	90,000	(20,000)		70,000	50,000	20,000	8,000	12,000	62,000
8	2	100,000	(20,000)		80,000	80,000	0	0	0	80,000
9	3	60,000	(22,000)		38,000	48,000	(10,000)	(4,000)	(6,000)	42,000
10	4	60,000	(24,000)		36,000	28,800	7,200	2,880	4,320	33,120
11	5	60,000	(26,000)		34,000	28,800	5,200	2,080	3,120	31,920
12	6	40,000	(28,000)	0	12,000	14,400	(2,400)	(960)	(1,440)	12,960
13						250,000				

17.17 (a) Find BV_3 after 3 years of MACRS depreciation.
       $BV_3 = 40,000 – 8000 – 12,800 – 7680 = \$11,520$

       (b) Sell asset for $BV_3 = \$11,520$.

Year	CFBT	P or S	D	TI	Taxes	CFAT
0	-	–40,000	-	-	-	-$40,000
1	20,000		8,000	12,000	4,800	15,200
2	20,000		12,800	7,200	2,880	17,120
3	20,000	11,520	7,680	12,320	4,928	26,592

17.18 (a) For SL depreciation with n = 3 years,
       $D_t = \$50,000$ per year
       Taxes = TI(0.35)

t	CFBT	D_t	TI	Taxes
1-3	$80,000	$50,000	$30,000	$10,500

$PW_{tax} = 10,500(P/A,15\%,3)$
$= 10,500(2.2832)$
$= \$23,974$

For MACRS depreciation, use Table 16.2 rates.

t	CFBT	d_t	D_t	TI	Taxes
1	$80,000	33.33%	$49,995	$30,005	$10,502
2	80,000	44.45	66,675	13,325	4,664
3	80,000	14.81	22,215	57,785	20,225
4	0	7.41	11,115	-11,115	-3,890

$PW_{tax} = 10,502(P/F,15\%,1) + \ldots - 3890(P/F,15\%,4)$
$= \$23,733$

MACRS has only a slightly lower PW_{tax} value.

(b) Total taxes: SL is 3(10,500) = $31,500
MACRS is 10,502 + ... - 3890 = $31,501   (rounding error)

17.19 (a) <u>U.S. Asset - MACRS</u>
For each year, use D_t for MACRS with n = 5.
$TI = CFBT - D_t = 65,000 - D_t$
$Taxes = 0.40\ TI$

t	d_t	D_t	TI	Taxes
1	0.20	$50,000	$15,000	$6000
2	0.32	80,000	-15,000	-6000
3	0.192	48,000	17,000	6800
4	0.1152	28,800	36,200	14,480
5	0.1152	28,800	36,200	14,480
6	0.0576	14,400	50,600	20,240
				$56,000

$PW_{tax} = 6000(P/F,12\%,1) - \ldots + 20,240(P/F,12\%,6)$
$= \$33,086$

<u>Italian Asset - Classical SL</u>
Calculate SL depreciation and TI for all 5 years.
$D_t = (250,000 - 25,000)/5 = \$45,000$
$TI = \$65,000 - 45,000 = \$20,000$

t	D_t	TI	Taxes
1	$45,000	20,000	$8000
2	45,000	20,000	8000
3	45,000	20,000	8000
4	45,000	20,000	8000
5	45,000	20,000	8000
6	0	65,000	26,000
			$66,000

PW_{tax} = 8000(P/A,12%,5) + 26,000(P/F,12%,6) = $42,010

As expected, MACRS has a smaller PW_{tax}

(b) Total taxes are $56,000 for MACRS and $66,000 for classical SL. The SL depreciation has S = $25,000, so a total of (25,000)(0.4) more in taxes is paid. Also, there are no taxes included on the depreciation recapture of $25,000 in year 6. This generates the $10,000 difference in total taxes.

17.20 Find the difference between PW of CFBT and CFAT.

t	CFBT	d_t	D_t	TI	Taxes	CFAT
1	$10,000	0.20	$1,800	$8,200	$3,280	$6720
2	10,000	0.32	2,880	7,120	2,848	7152
3	10,000	0.192	1,728	8,272	3,309	6691
4	10,000	0.1152	1,037	8,963	3,585	6415
5	5,000	0.1152	1,037	3,963	1,585	3415
6	5,000	0.0576	518	4,482	1,793	3207

PW_{CFBT} = 10,000(P/A,10%,4) + 5000(P/A,10%,2)(P/F,10%,4) = $37,626
PW_{CFAT} = 6720(P/F,10%,1) + ... + 3207(P/F,10%,6) = $25,359

Cash flow lost to taxes is $12,267 in PW dollars.

17.21 (a)
$$PW_{TS} = \sum_{t=1}^{t=n} \text{(tax savings in year t)}(P/F,i,t)$$

Select the method that maximizes PW_{TS} which is the opposite of minimizing the PW_{tax} value.

(b) $TS_t = D_t(0.35)$

t	d_t	D_t	TS_t
1	0.3333	$14,999	$5,250
2	0.4445	20,002	7,001
3	0.1481	6,665	2,333
4	0.0741	3,334	1,167

PW_{TS} = 5,250(P/F,8%,1) + ...+ 1,167(P/F,8%,4) = $13,573

17.22

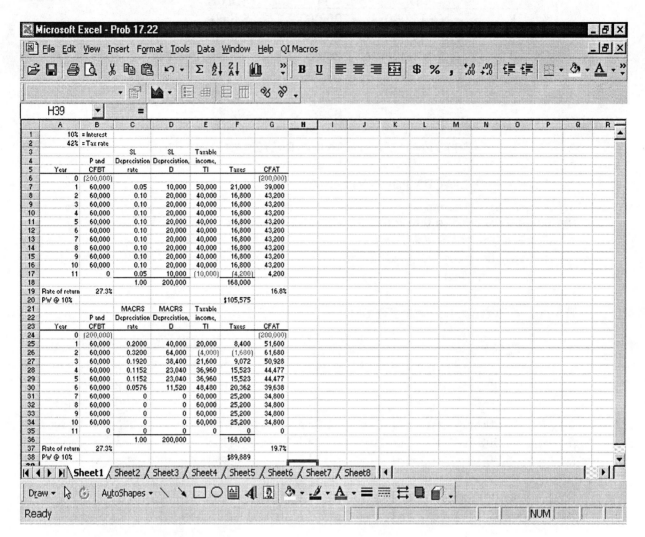

(a and b)

	SL	MACRS
i* of CFBT	27.3%	27.3%
i* of CFAT	16.8%	19.7%

MACRS raises the after tax i* because of accelerated depreciation.

(c) Select MACRS with $PW_{tax} = \$89,889$

17.23   <u>MACRS with n = 5 years</u>

t	CFBT	D_t	TI	Taxes
1	$4,000	$2,000	$2,000	$ 700
2	4,000	3,200	800	280
3	4,000	1,920	2,080	728
4	4,000	1,152	2,848	997
5	4,000	1,152	2,848	997
6	4,000	576	3,424	1,198

MACRS PW_{tax} = $2482

<u>Straight Line with n = 8 years</u>

t	CFBT	D_t	TI	Taxes
1	$4,000	$ 625	$3,375	$1,181
2-8	4,000	1,250	2,750	963
9	4,000	625	3,375	1,181

SL alternative PW_{tax} = 1,181[(P/F,20%,1) + (P/F,20%,9)]
                            + 963(P/A,20%,7)(P/F,20%,1)
                        = $4106

Select the 5-year MACRS method.

17.24   1. Selling price = 0.4(150,000) = $60,000
         BV_4 = 150,000(1 – 0.6876) = $46,860
         DR = SP – BV_4 = $13,140
         Taxes = DR(T_e) = 13,140(0.35) = $4599

      2. CG = $10,000
         DR = 0.3333(100,000) = $33,330
         TI = $43,330
         Taxes = $15,166

      3. Land does not depreciate.
         CG = TI = 0.1(1.8 million) = $180,000
         Taxes = 180,000(0.35) = $63,000

      4. CL = 5000 – 500 = $4500
         TI = $–4500
         Taxes = 0.35(–4500) = $–1575

5. $DR = TI = 2000$
   Taxes = $2000(0.35) = \$700$

17.25 Use MACRS rates for n = 5

$BV_2 = 40{,}000 - 0.52(40{,}000) = \$19{,}200$

There is depreciation recapture (DR)
$DR = 28{,}000 - 19{,}200 = \$8800$

$TI = GI - E - D + DR$
$ = 20{,}000 - 3000 - 0.32(40{,}000) + 8800 = \$13{,}000$

Taxes = $13{,}000(0.30) = \$3900$

17.26 Land: CG = $25,000
Building: CL = $55,000
Asset 1: DR = $18{,}500 - 15{,}500 = \$3000$
Asset 2: DR = $10{,}000 - 5{,}000 = \$5{,}000$
 CG = $10{,}500 - 10{,}000 = \$500$

17.27 In year 4, a DR amount of $20,000 is additional TI.
In $10,000 units, at the time of sale in year 4:

Year	GI	S	E	D	TI	Taxes	CFAT
4	$10	$2	$5	$1.482	$5.518	2.2072	$4.7928

$CFAT = GI - E + S - \text{taxes}$
$ = 10 - 5 + 2 - 2.2072$
$ = \$4.7928 \qquad (\$47{,}928)$

CFAT decreased from $55,930 in Prob.17.14(b).

17.28

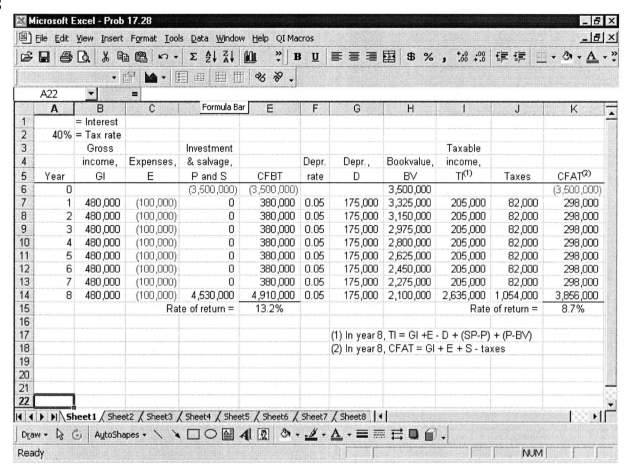

(a) CFAT is in column K.

(b) Before-tax ROR = 13.2% (cell E15).
    After-tax ROR = 8.7%   (cell K15)

17.29 SL depreciation = 10,000/5 = $2,000; CFBT = GI - E

Year	P and S	CFBT	D	TI	Taxes	CFAT
0	$–10,000	-	-	-	-	$–10,000
1-5		$5000	$2000	$3000	$1440	3560
6		5000	0	5000	2400	2600
6	3075	-	-	3075	1476	1599
						$11,999

17.30 (a)

Year	P and S	CFBT	D	TI	Taxes	CFAT
0	$-10,000	-	-	-	-	$-10,000
1	-	$5000	$2000	$3000	$1440	3560
2	-	5000	3200	1800	864	4136
3	-	5000	1920	3080	1478	3522
4	-	5000	1152	3848	1847	3153
5	-	5000	1152	3848	1847	3153
6	3075	5000	576	7499*	3600	4475
						$ 11,999

*TI = (5000 – 576 + 3075) = $7499

(b) Total CFAT = $11,999 is the same as for classical SL; only the timing is different.

17.31 Straight line depreciation

$D_t = \frac{45,000 - 3000}{5} = \$8400$

TI = 15,000 – 8400 = $6600
Taxes = 6600(0.5) = $3300
No depreciation recapture is involved.

$PW_{tax} = 3300(P/A, 18\%, 5) = \$10,320$

DDB-to-SL switch

TI = 15,000 – D_t
Taxes = TI(0.50)

The depreciation schedule was determined in Problem 16A.4.

t	CFBT	D_t	Method	TI	Taxes
1	$15,000	18,000	DDB	$-3000	$-1500
2	15,000	10,800	DDB	4200	2100
3	15,000	6480	DDB	8520	4260
4	15,000	3888	DDB	11,112	5556
5	15,000	2832	SL	12,168	6084

No depreciation recapture is involved.

$PW_{tax} = -1500(P/F, 18\%, 1) + \ldots + 6084(P/F, 18\%, 5)$
$= \$8355$

Switching gives a $1965 lower PW_{tax} value.

17.32 Effective tax rate = 0.042 + (1 − 0.042) (0.34)
= 0.36772

$$\text{Before-tax ROR} = \frac{0.07}{1-0.36772} = 0.111$$

Here, 11.1% before-taxes is equivalent to 7% after-taxes.

17.33
$$0.12 = \frac{0.08}{1-\text{tax rate}}$$
1-tax rate = 0.667
Tax rate = 0.333   (33.3%)

17.34   Small company

Use Equation [17.17].

After-tax ROR = 0.21(1−0.34) = 0.1386   (13.86%)
Conclusion: Accept at MARR = 12%

           Large corporation

After-tax ROR = 0.21(1−0.48) = 0.1092   (10.92%)
Conclusion: Reject at MARR = 12%

17.35

	Cell	Before-tax	Cell	After-tax
PW:	B12	=-PV(14%,5,75000,15000)-200000	C12	=NPV(9%,C6:C10)+C5
AW:	B13	=PMT(14%,5,-B12)	C13	=PMT(9%,5,-C12)
IRR:	B14	=IRR(B5:B10)	C14	=IRR(C5:C10)

17.36 (a) Solution by Computer

Use the spreadsheet format of Figure 17-3b with a column for BV added. The two spreadsheets are shown below.

Results: PW_A = $-451 and PW_B = $835. Select machine B.

(b) Solution by hand

Develop tables similar to the 2 spreadsheets.

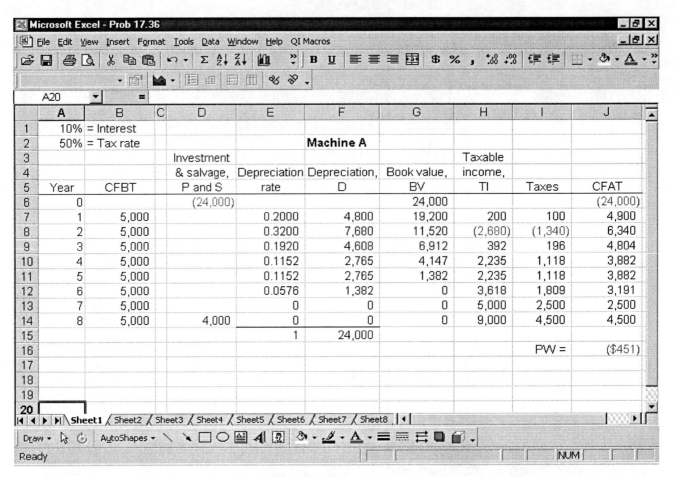

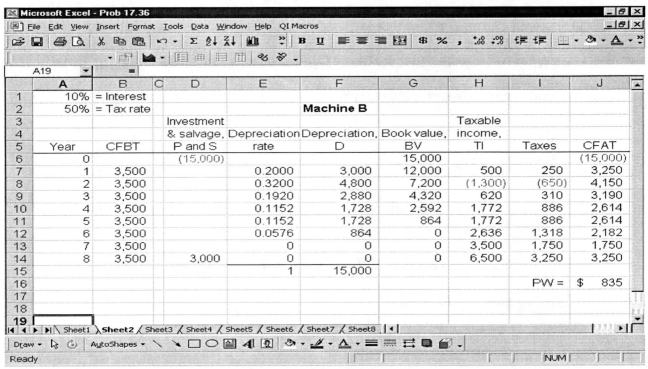

Chapter 17

17.37 (a) $PW_A = -15{,}000 - 3000(P/A,14\%,10) + 3000(P/F,14\%,10)$
$\quad\quad\quad = -15{,}000 - 3000(5.2161) + 3000(0.2697)$
$\quad\quad\quad = \$\text{-}29{,}839$

$PW_B = -22{,}000 - 1500(P/A,14\%,10) + 5000(P/F,14\%,10)$
$\quad\quad\quad = -22{,}000 - 1500(5.2161) + 5000(0.2697)$
$\quad\quad\quad = \$\text{-}28{,}476$

Select B with the lower PW value.

(b) All costs generate tax savings.

Machine A

Annual depreciation = $(15{,}000 - 3{,}000)/10 = \$1200$
Tax savings = $(AOC + D)0.5 = 4200(0.5) = \$2100$
CFAT = $-3000 + 2100 = \$\text{-}900$

$PW_A = -15{,}000 - 900(P/A,7\%,10) + 3000(P/F,7\%,10)$
$\quad\quad\quad = -15{,}000 - 900(7.0236) + 3000(0.5083)$
$\quad\quad\quad = \$\text{-}19{,}796$

Machine B

Annual depreciation = $\dfrac{22{,}000 - 5000}{10} = \$1700$
Tax savings = $(1500 + 1700)(0.50) = \$1600$
CFAT = $-1500 + 1600 = \$100$

$PW_B = -22{,}000 + 100(P/A,7\%,10) + 5000(P/F,7\%,10)$
$\quad\quad\quad = -22{,}000 + 100(7.0236) + 5000(0.5083)$
$\quad\quad\quad = \$\text{-}18{,}756$

Select machine B.

(c) MACRS with n = 5 and a DR in year 10, which is a tax, not a tax savings.
Tax savings = $(AOC + D)(0.5)$, years 1-6
CFAT = $-AOC$ + tax savings, years 1-10.

Machine A

Year 10 has a DR tax of $3{,}000(0.5) = \$1500$

Year	P and S	AOC	D	Tax savings	CFAT
0	$–15,000	-	-	-	$–15,000
1		$3000	$3000	$3000	0
2		3000	4800	3900	900
3		3000	2880	2940	-60
4		3000	1728	2364	-636
5		3000	1728	2364	-636
6		3000	864	1932	-1068
7		3000	0	1500	-1500
8		3000	0	1500	-1500
9		3000	0	1500	-1500
10		3000	0	1500	-1500
10	3000	-	-	–1500	1500

$PW_A = -15,000 + 0 + 900(P/F,7\%,2) + \ldots - 1,500(P/F,7\%,9)$
$= \$-18,536$

Machine B

Year 10 has a DR tax of $5,000(0.5) = \$2,500$

Year	P and S	AOC	D	Tax savings	CFAT
0	$–22,000	-	-	-	$–22,000
1		$1500	$4400	$2950	1450
2		1500	7040	4270	2770
3		1500	4224	2862	1362
4		1500	2534	2017	517
5		1500	2534	2017	517
6		1500	1268	1384	–116
7		1500	0	750	–750
8		1500	0	750	–750
9		1500	0	750	–750
10		1500	0	750	–750
10	5000	-	-	–2500	2500

$PW_B = -22,000 + 1450(P/F,7\%,1) + \ldots + 2500(P/F,7\%,10)$
$= \$-16,850$

Select machine B, as above.

17.38

Alternative A

Year	P and S	GI-E	D	TI	Taxes	CFAT
0	$-8000	-	-	-	-	$-8000
1		$3500	$2666	834	333	3167
2		3500	3556	-56	-22	3522
3		3500	1185	2315	926	2574
4	0	0	593	-593	-237	237

$PW_A = \$169$

Alternative B

Year	P and S	GI-E	D	TI	Taxes	CFAT
0	$-13,000	-	-	-	-	$-13,000
1		$5000	$4333	$667	$267	4733
2		5000	5779	-779	-311	5311
3		5000	1925	3075	1230	3770
4		0	963	-963	-385	385
	2000	-	-	2000	800	1200

$PW_B = \$94$

Select alternative A.

17.39

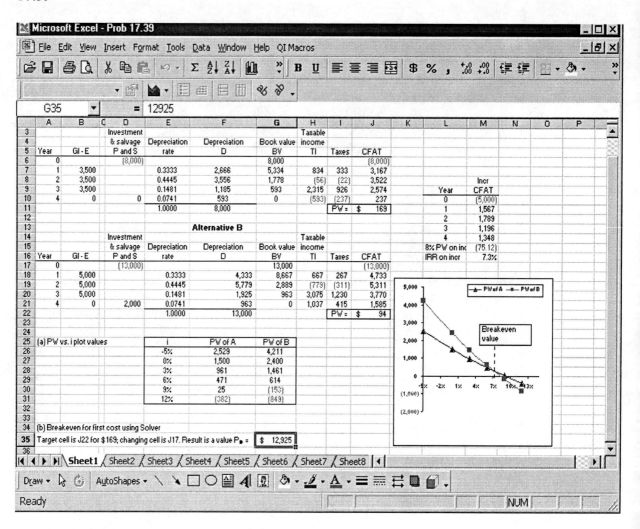

17.40

<u>Cooker A</u>

Depreciation = 15,000/3 = $5000

For years 1 to 3:

$$TI = 6600 - 5000 = \$1600$$
$$Taxes = 1600(0.35) = \$560$$
$$CFAT = 6600 - 560 = \$6040$$

$AW_A = -15,000(A/P,6\%,3) + 6040$
$ = -15,000(0.37411) + 6040$
$ = \$428$

<u>Cooker B</u>

Depreciation = 25,000/5 = $5000

For years 1 to 5:

$$TI = 9000 - 5000 = \$4000$$
$$Taxes = 4000(0.35) = \$1400$$
$$CFAT = 9000 - 1400 = \$7600$$

For year 5 only:

$$DR = 25,000(0.10) = \$2500$$
$$DR\ taxes = 2500(0.35) = \$875$$

$$\begin{aligned}AW_B &= -25,000(A/P,6\%,5) + 7600 + (2500-875)(A/F,6\%,5) \\ &= -25,000(0.23740) + 7600 + 1625(0.17740) \\ &= \$1953\end{aligned}$$

Select cooker B.

17.41 (a) Classical SL with n = 5 year recovery period.

$$\text{Annual depreciation} = \frac{2,500 - 0}{5} = \$500$$

<u>Year 1</u>

$$Taxes = (1,500 - 500)(0.45) = \$450$$
$$CFAT = 1,500 - 450 = \$1,050$$

<u>Years 2-5</u>

$$Taxes = (300 - 500)(0.45) = \$-90$$
$$CFAT = 300 - (-90) = \$390$$

The rate of return relation is:

$$0 = -2,500 + 1,050(P/F,i^*,1) + 390\ (P/A,i^*,4)(P/F,i^*,1)$$

$$i^* = 1.76\% \qquad\qquad \text{(IRR function)}$$

(b) Use MACRS with n = 5 year recovery period.

Year	P	GI-E	D	TI	Taxes	CFAT
0	$–2,500	-	-	-	-	-$2,500
1		$1,500	$500	$1,000	$450	1,050
2		300	800	-500	-225	525
3		300	480	-180	-81	381
4		300	288	12	5	295
5		300	288	12	5	295
6		300	144	156	70	230

The ROR relation is:
$$0 = -2500 + 1050(P/F,i^*,1) + ... + 230(P/F,i^*,6)$$
$$i^* = 4.17\% \qquad \text{(IRR function)}$$

The use of MACRS increases i^* from the 1.76% for classical SL depreciation.

17.42 For a 10% after-tax return, solve for n in a PW relation.

$$-78,000 + 15,000(P/A,10\%,n) = 0$$
$$(P/A,10\%,n) = 5.2$$

n = 7.7 years (interpolation)

Keep the inspection equipment for 2.7 more years.

17.43 (a) Get the CFAT values from Problem 17.37(b) for SL depreciation. Use a spreadsheet to find the incremental ROR (column D) and to determine the PW of incremental CFAT versus incremental i values (columns E and F) for the chart.

The $\in i^* = 9.75\%$ can also be found using the PW relation:

$$0 = -7000 + 1000(P/A,i^*,9) + 3000(P/F,i^*,10)$$

If MARR < 9.75%, select B, otherwise select A.

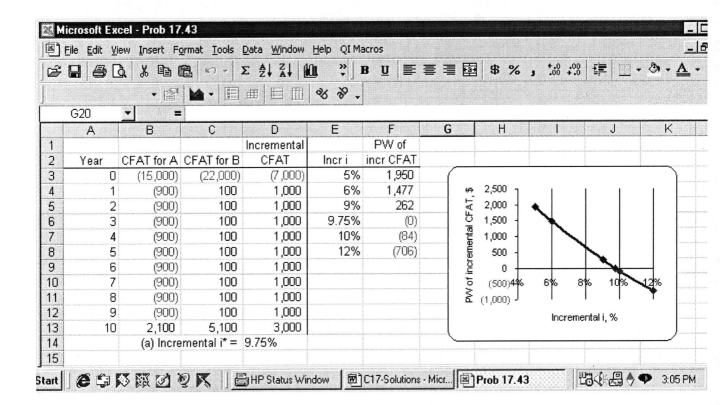

(b) Use the PW vs. incremental i plot to select between A and B at each MARR value.

MARR	Select
6%	B
9	B
10	A
11	A

17.44 (a) The equation to determine the first cost P is

$$0 = -P + (CFBT - \text{taxes})(P/A, 20\%, 5)$$
$$= -P + [20{,}000 - (20{,}000 - P/5)(0.40)](P/A, 20\%, 5)$$
$$= -P + [12{,}000 + 0.08P](2.9906)$$
$$= -P + 35{,}887 + 0.23925P$$
$$P = \$47{,}173$$

(b) The equation to find the S value is:
$$0 = -50{,}000 + \{20{,}000 - [20{,}000 - (50{,}000 - S)/5](0.40)\}$$
$$(P/A, 20\%, 5) + S(P/F, 20\%, 5)$$

$$0 = -50,000 + \{20,000 - [20,000 - (10,000 - S/5)](0.4)\}(2.9906) + 0.4019S$$
$$= -50,000 + \{20,000 - [4000 + 0.08S]\}(2.9906) + 0.4019S$$
$$= -50,000 + \{16,000 - 0.08S\}(2.9906) + 0.4019S$$
$$= -2150 - 0.23925S + 0.4019S$$

$$S = 2150/0.16265$$
$$= \$13,218$$

17.45 (a) The lives are established at 8 (remaining) years for the defender and 8 years for the challenger.

Defender

Annual depreciation = $\dfrac{28,000 - 2000}{13} = 2000$

Annual tax savings = $(2000 + 1200)(0.48) = \$1536$

$AW_D = -18,000(A/P,8\%,8) + 2000(A/F,8\%,8) - 1200 + (2000 + 1200)(.48)$
$= -18,000(0.17401) + 2000(0.09401) - 1200 + 1536$
$= \$-2608$

Challenger

DR from sale of D = Market value $- BV_5$
$= 18,000 - [28,000 - 5(2000)] = 0$

Challenger annual depreciation = $\dfrac{15,000 - 3000}{8} = \$1500$

Annual tax saving = $(1,500 + 1,500)(0.48) = \$1440$

Challenger DR, year 8 = $3000 - 3000 = 0$

$AW_C = -15,000(A/P,8\%,8) + 3000(A/F,8\%,8) - 1500 + 1440$
$= -15,000(0.17401) + 3000(0.09401) - 60$
$= \$-2388$

Select the challenger

(b) $AW_D = -18,000(A/P,15\%,8) + 2000(A/F,15\%,8) - 1200$
$= -18,000(0.22285) + 2000(0.07285) - 1200$
$= \$-5066$

$AW_C = -15,000(A/P,15\%,8) + 3000(A/F,15\%,8) - 1500$
$= -15,000(0.22285) + 3000(0.07285) - 1500$
$= \$-4624$

Select the challenger. The before-tax decision is the same as the after-tax decision.

17.46 Challenger

DR for C, year 8 = 8000 – 3000 = $5000

DR tax = $5000(0.48) = $2400

AW_C = –15,000(A/P,8%,8) + 3000(A/F,8%,8) – 1500 + 1440 – 2400(A/F,8%,8)
     = –15,000(0.17401) + 3000(0.09401) – 60 – 2400(0.09401)
     = $–2614

From Prob. 17.45(a), AW_D = $–2608
Select defender now (marginally).

17.47 Study period is 3 years.

Option	Defender	Challenger
1	2 years	1 year
2	1	2
3	0	3

Defender

AW_D = $–250,000   for 1 or 2 years

Challenger

No tax effect if (defender) contract is cancelled. Calculate CFAT for 1, 2, and 3 years of ownership.

Year	E	d	D	BV	SP	DR/CL	TI	Tax Savings	CFAT
0	-	-	-	$500,000	-	-	-	-	$–500,000
1	$80,000	0.3333	$166,650	333,350	$333,333	$–17^{CL}	$–246,667	$–123,334	376,667
2	80,000	0.4445	222,250	111,100	166,667	55,567^{DR}	–246,683	–123,342	210,009
3	80,000	0.1481	74,050	37,050	0	37,050^{CL}	–191,100	–95,550	15,550

TI = –E – D + DR – CL

Year 1:   TI = –80,000 – 166,650 – 17 = $–246,667
Year 2:   TI = –80,000 – 222,250 + 55,567 = $–246,683
Year 3:   TI = –80,000 – 74,050 – 37,050 = $–191,100

CFAT = –E + SP – taxes   where negative taxes are a tax savings

Year 1:   –80,000 + 333,333 – (–123,334) = 376,667

$$AW_{C1} = -500,000(A/P,7\%,1) + 376,667$$
$$= -500,000\,(1.07) + 376,667$$
$$= \$\!-158,333$$

$$AW_{C2} = -500,000(A/P,7\%,2) + [376,667(P/F,7\%,1) + 210,009(P/F,7\%,2)](A/P,7\%,2)$$
$$= -500,000(0.55309) + [376,667(0.9346) + 210,009(0.8734)](0.55309)$$
$$= \$\!+19,610$$

$$AW_{C3} = -500,000(A/P,7\%,3) + [(376,667)(P/F,7\%,1) + 210,009(P/F,7\%,2)$$
$$+ 15,550(P/F,7\%,3)](A/P,7\%,3)$$
$$= [-500,000 + 376,667(0.9346) + 210,009(0.8734) + 15,550(0.8163)](0.38105)$$
$$= \$18,347$$

Summary of cost/year and AW

Option	Year 1	Year 2	Year 3	AW
1	$-250,000	$-250,000	$-158,333	$-221,484
2	-250,000	19,610	19,610	-76,406
3	18,347	18,347	18,347	18,347

Replace now with the challenger.

17.48 (a) Study period is set at 5 years. The only option is the defender for 5 years and the challenger for 5 years.

Defender

First cost = Sale + Upgrade
= 15,000 + 9000
= $24,000

Upgrade SL depreciation = $3000 year   (years 1-3 only)
AOC, years 1-5:   = $6000
Tax saving, years 1-3:   = (6000 + 3000)(0.4)
                          = $3600
Tax savings, year 4-5:   = 6000(0.4) = $2,400
Actual cost, years 1-3:   = 6000 - 3600 = $2400
Actual cost, years 4-5:   = 6000 - 2400 = $3600

$$AW_D = -24,000(A/P,12\%,5) - 2400 - 1200(F/A,12\%,2)(A/F,12\%,5)$$
$$= -24,000(0.27741) - 2400 - 1200(2.12)(0.15741)$$
$$= \$\!-9458$$

Challenger

DR on defender = $15,000
DR tax = $6000
First cost + DR tax = $46,000
Depreciation = 40,000/5 = $8,000
Expenses = $7,000 (years 1-5)
Tax saving = (8000 + 7000)(0.4) = $6,000
Actual AOC = 7000 – 6000 = $1000 (years 1-5)

AW_C = –46,000(A/P,12%,5) – 6000
= –46,000(0.27741) – 1000
= $–13,761

Retain the defender since the AW of cost is smaller.

(b) AW_C will become less costly, but the revenue from the challenger's sale between $2000 to $4000 will be reduced by the tax on DR in year 5.

17.49

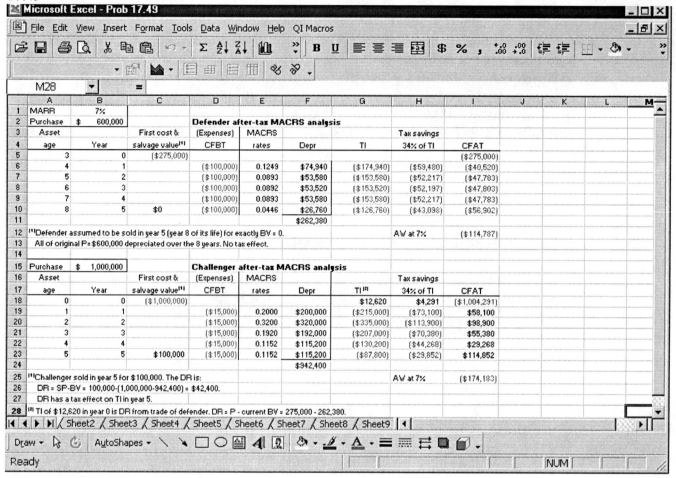

Still select the defender but with a larger AW advantage.

17.50 (a) The EVA shows the monetary worth added to a corporation by an alternative.

(b) The EVA estimates can be used directly in public reports (e.g., to stockholders). EVA shows worth contribution, not just CFAT.

17.51 (a) Take TI, taxes and D from Example 17.3. Use $i = 0.10$ and $T_e = 0.35$.

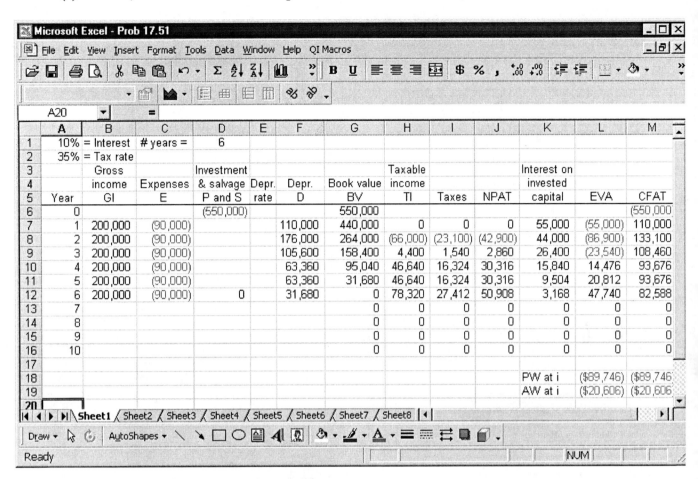

$NPAT = TI(1 - 0.35)$
$EVA = NPAT -$ interest of invested capital

(b) The spreadsheet shows that the two AW values are equal. Solution by hand is as follows:

$AW_{EVA} = [-55,000(P/F,10\%,1) + \ldots + 47,740(P/F,10\%,6)](A/P,10\%,6)$
$= [-55,000(0.9091) + \ldots + 47,740(0.5645)](0.22961)$
$= -89,746(0.22961)$
$= \$-20,606$

$AW_{CFAT} = [-550,000 + 110,000(P/F,10\%,1) + \ldots + 82,588(P/F,10\%,6)](A/P,10\%,6)$
$= [-550,000 + 110,000(0.9091) + \ldots + 82,588(0.5645)](0.22961)$
$= -89,746(0.22961)$
$= \$-20,606$

17.52 (a) Column L shows the EVA each year. Use Eq. [17.18] to calculate EVA.

(b) The AW_{EVA} = $338,000 is calculated on the spreadsheet.

Note: The CFAT and AW_{CFAT} values are also shown on the spreadsheet.

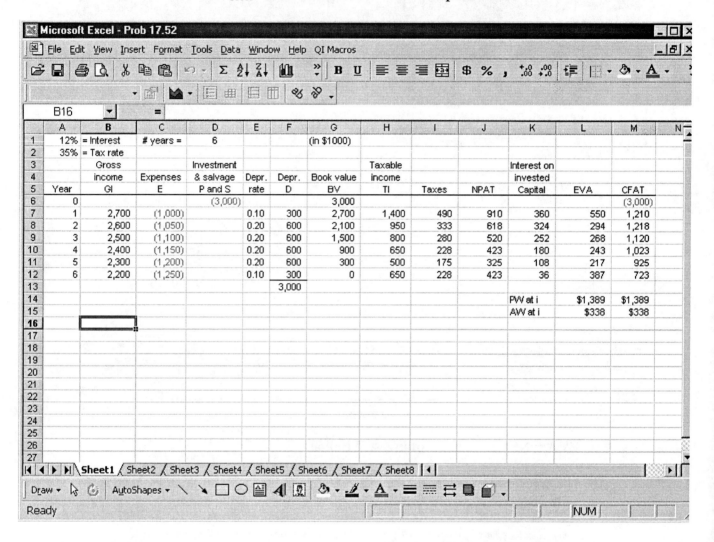

17.53 The spreadsheet shows EVA for both analyzers. Select analyzer 2 with the larger AW of EVA. This is the same decision as that reached using an AW of CFAT analysis.

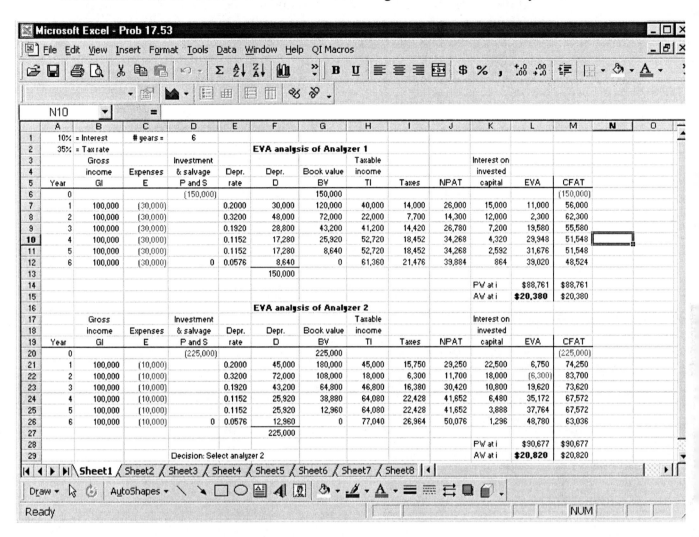

Case Study Solution

1. Set up on the next two spreadsheets. The 90% debt option has the largest PW at 10%. As mentioned in the chapter, the largest D-E financing option will always offer the largest return on the invested equity capital. But, too high D-E mixes are risky.

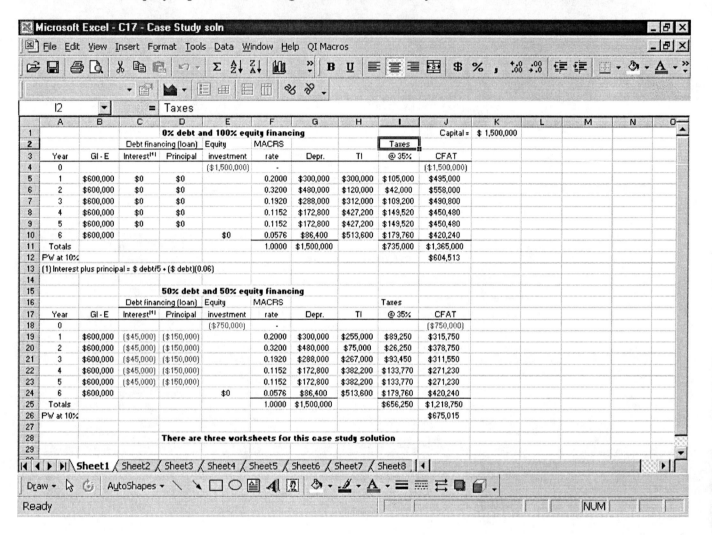

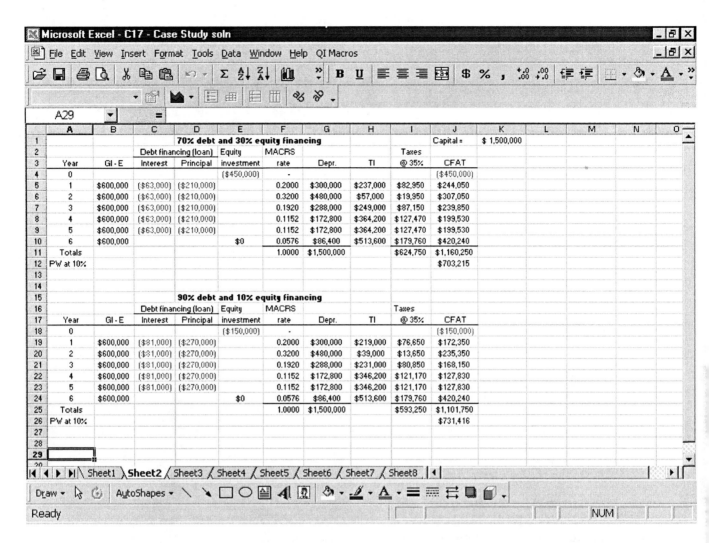

2. Subtract 2 different equity CFAT totals.
   For 30% and 10%:
   (1,160,250 – 1,101,750) = $58,500
   Divide by 2 to get the change per 10% equity.
   58,500/2 = $29,250
   Conclusion: Total CFAT increases by $29,250 for each 10% increase in equity financing.

3. This happens because less of the Young Brothers own funds are committed to the
   Portland branch the larger the loan principal.

4. The best estimates of annual EVA are shown in column M.
   The equivalent AW = $113,342.

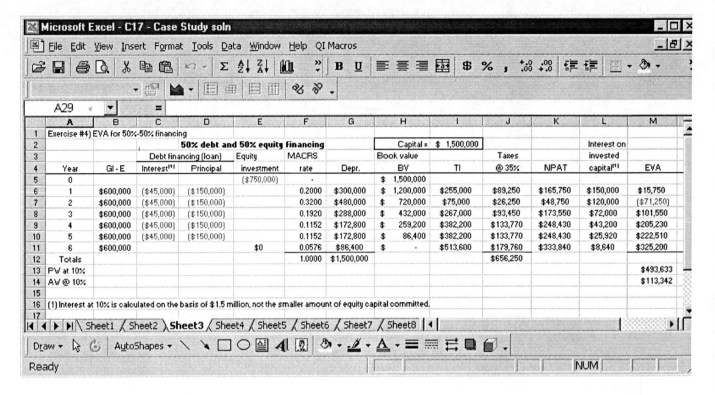

Equations used to determine the EVA:

EVA = NPAT − interest on invested capital
NPAT = TI − taxes
(Interest on invested capital)$_t$ = i(BV in the previous year)
   = $0.10(BV_{t-1})$

Note: BV on the entire $1.5 million in depreciable assets is used to determine the interest on invested capital.

Chapter 18 – Formalized Sensitivity Analysis and Expected Value Decisions
Solutions to end of chapter exercises

Problems

18.1

<u>10 tons/day</u>

$$PW = -62{,}000 + 1500(P/F,10\%,8) - 0.50(10)(200)(P/A,10\%,8) - 4(8)(200)(P/A,10\%,8)$$
$$= -62{,}000 + 1500(0.4665) - 7400(5.3349)$$
$$= \$-100{,}779$$

<u>20 tons/day</u>

$$PW = -62{,}000 + 1500(P/F,10\%,8) - 0.50(20)(200)(P/A,10\%,8)$$
$$\quad - 8(8)(200)(P/A,10\%,8)$$
$$= -62{,}000 + 1500(0.4665) - 14{,}800(5.3349)$$
$$= \$-140{,}257$$

<u>30 tons/day</u>

Overtime hours required = (30/20)8 – 8 = 4.0 hours

$$PW = -62{,}000 + 1500(P/F,10\%,8) - 0.50(30)(200)(P/A,10\%,8)$$
$$\quad - [8(8)(200) + 4.0(16)(200)](P/A,10\%,8)$$
$$= -62{,}000 + 1500(0.4665) - 28{,}600(5.3349)$$
$$= \$-213{,}878$$

18.2 Joe: $PW = -77{,}000 + 10{,}000(P/F,8\%,6) + 10{,}000(P/A,8\%,6)$
$\quad = -77{,}000 + 10{,}000(0.6302) + 10{,}000(4.6229)$
$\quad = \$-24{,}469$

Jane: $PW = -77{,}000 + 10{,}000(P/F,8\%,6) + 14{,}000(P/A,8\%,6)$
$\quad = -77{,}000 + 10{,}000(0.6302) + 14{,}000(4.6229)$
$\quad = \$-5977$

Carlos: $PW = -77{,}000 + 10{,}000(P/F,8\%,6) + 18{,}000(P/A,8\%,6)$
$\quad = -77{,}000 + 10{,}000(0.6302) + 18{,}000(4.6229)$
$\quad = \$12{,}514$

Only the $18,000 revenue estimate of Carlos favors the investment.

18.3 Set up the spreadsheets for income estimates of $10,000, 14,000 and 18,000 and calculate the PW at 8(1-0.35) = 5.2%. The $18,000 revenue estimate is the only one with PW > 0.

Row	A	B	C	D	E MACRS	F Taxable	G	H
1	Joe: $10,000 = Revenue estimate				MACRS	Taxable		
2	Year	Revenue	Expenses	P and S	Depreciation	income	Taxes	CFAT
3	0			-77000				$ (77,000)
4	1	10000	2000		$ 15,400	$ (7,400)	$ (2,590)	$ 10,590
5	2	10000	2000		$ 24,640	$ (16,640)	$ (5,824)	$ 13,824
6	3	10000	2000		$ 14,784	$ (6,784)	$ (2,374)	$ 10,374
7	4	10000	2000		$ 8,870	$ (870)	$ (305)	$ 8,305
8	5	10000	2000		$ 8,870	$ (870)	$ (305)	$ 8,305
9	6	10000	2000	10000	$ 4,435	$ (6,435)	$ (2,252)	$ 20,252
11	TI = B - C - D							
12	CFAT = B - C + D - G				PW of CFAT =	$ (17,365)		
13	Depr. Recapture in year 6 is $10,000							
15	Jane: $14,000 = Revenue estimate				MACRS	Taxable		
16	Year	Revenue	Expenses	P and S	Depreciation	income	Taxes	CFAT
17	0			-77000				$ (77,000)
18	1	14000	2000		$ 15,400	$ (3,400)	$ (1,190)	$ 13,190
19	2	14000	2000		$ 24,640	$ (12,640)	$ (4,424)	$ 16,424
20	3	14000	2000		$ 14,784	$ (2,784)	$ (974)	$ 12,974
21	4	14000	2000		$ 8,870	$ 3,130	$ 1,095	$ 10,905
22	5	14000	2000		$ 8,870	$ 3,130	$ 1,095	$ 10,905
23	6	14000	2000	10000	$ 4,435	$ (2,435)	$ (852)	$ 22,852
25					PW of CFAT =	$ (4,252)		
27	Carlos: $18,000 = Revenue estimate				MACRS	Taxable		
28	Year	Revenue	Expenses	P and S	Depreciation	income	Taxes	CFAT
29	0			-77000				$ (77,000)
30	1	18000	2000		$ 15,400	$ 600	$ 210	$ 15,790
31	2	18000	2000		$ 24,640	$ (8,640)	$ (3,024)	$ 19,024
32	3	18000	2000		$ 14,784	$ 1,216	$ 426	$ 15,574
33	4	18000	2000		$ 8,870	$ 7,130	$ 2,495	$ 13,505
34	5	18000	2000		$ 8,870	$ 7,130	$ 2,495	$ 13,505
35	6	18000	2000	10000	$ 4,435	$ 1,565	$ 548	$ 25,452
37					PW of CFAT =	$ 8,861		

18.4

$$PW_{Build} = -80{,}000 - 70(1000) + 120{,}000(P/F,20\%,3)$$
$$= -150{,}000 + 120{,}000(0.5787)$$
$$= \$-80{,}556$$

$$PW_{Lease} = -(2.5)(12)(1000) - (2.50)(12)(1000)(P/A,20\%,2)$$
$$= -18{,}000 - 18{,}000(1.5278)$$
$$= \$-75{,}834$$

The company should lease the space.

New construction cost = 70(0.90) = $63 and lease at $2.75

$$PW_{Build} = -80{,}000 - 63(1000) + 120{,}000(P/F,20\%,3)$$
$$= -143{,}000 + 120{,}000(0.5787)$$
$$= \$-73{,}556$$

$$PW_{Lease} = -2.75(12)(1000)[1 + (P/A,20\%,2)]$$
$$= -15{,}000(2.5278)$$
$$= \$-83{,}417$$

Select build, the decision is sensitive.

18.5 Calculate i^* for G = $1500, 2000 and 2500. Other gradient values can be used. All $ values are in $1000.
(a and b) Solution by Hand and Computer provide the same answers.

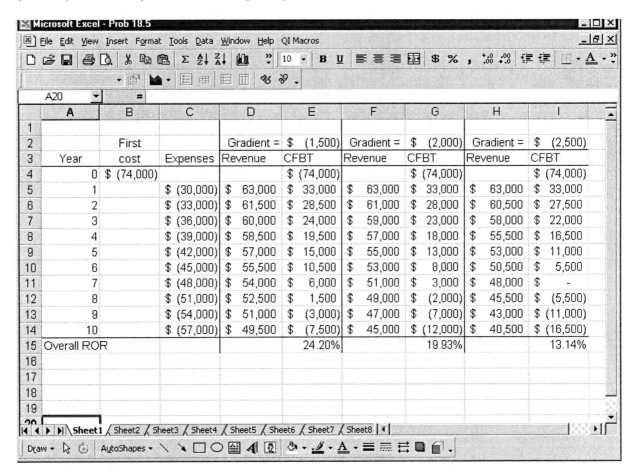

For MARR = 18%, the decision does change from YES for G = $1500 and $2000, to NO for G = $2500.

18.6 (a) The AW relations are:

$$AW_1 = -10{,}000(A/P,i,8) - 600 - 100(A/F,i,8) - 1750(P/F,i,4)(A/P,i,8)$$
$$AW_2 = -17{,}000(A/P,i,12) - 150 - 300(A/F,i,12) - 3000(P/F,i,6)(A/P,i,12)$$

(b) Spreadsheet analysis requires that the PW value is first determined using the NPV function over the LCM of 24 years, then converted to AW values by the PMT function.

Calculate AW for each MARR value. The decision is sensitive; it changes at 6%.

MARR	AW_1	AW_2	Select
4%	$–2318	$–2234	2
6	–2444	–2448	1
8	–2573	–2673	1

18.7 (a) Breakeven number of vacation days per year is x.

$$AW_{cabin} = -130{,}000(A/P,10\%,10) + 145{,}000(A/F,10\%,10) - 1500 + 150x - \frac{50}{30}(1.20)x$$

$$AW_{trailer} = -75{,}000(A/P,10\%,10) + 20{,}000(A/F,10\%,10) - 1{,}750 + 125x - \frac{300}{30(0.6)}(1.20)x$$

$$AW_{cabin} = AW_{trailer}$$

$$-130{,}000(0.16275) + 145{,}000(0.06275) - 1500 + 148x$$
$$= -75{,}000(0.16275) + 20{,}000(0.06275) - 1750 + 105x$$

$$-13{,}558.75 + 148x = -12{,}701.25 + 105x$$

$$43x = 857.5$$

$$x = 19.94 \text{ days} \quad \text{(Use } x = 20 \text{ days per year)}$$

(b) AW sensitivity analysis is performed for 12, 16, 20, 24, and 28 days.

$$AW_{cabin} = -13{,}558.75 + 148x$$
$$AW_{trailer} = -12{,}701.25 + 105x$$

Days, x	AW_{cabin}	$AW_{trailer}$	Selected
12	$-11,783	$-11,441	Trailer
16	-11,191	-11,021	Trailer
20	-10,599	-10,601	Cabin
24	-10,007	-10,181	Cabin
28	- 9415	- 9761	Cabin

Each pair of AW values are close to each other, especially for x = 20, which is the breakeven point.

(c) The trailer alternative. Select the alternative with the lower variable cost, since the variable term is positive, not a cost.

18.8 (a and b) Bond interest = b(50,000)/4 = $12,500(b), where b = 5%, 7%, and 9%.
Use trial and error (a) or the IRR function (b) to find i* in the PW relation:

$$0 = -42,000 + (12,500b)(P/A,i^*,60) - 50,000(P/F,i^*,60)$$

Rate, b	Interest per quarter	i*/quarter	Nominal i* per year
5%	$625	1.67%	6.68%
7%	875	2.24	8.96
9%	1125	2.80	11.20

18.9 <u>6 years</u>
PW = –30,000 + 3500(P/A,8%,6) + 25,000(P/F,8%,6)
= –30,000 + 3500(4.6229) + 25,000(0.6302)
= $1935

<u>10 years</u>
PW = –30,000 + 3500(P/A,8%,10) + 15,000(P/F,8%,10)
= –30,000 + 3500(6.7101) + 15,000)0.4632)
= $433

<u>12 years</u>
PW = –30,000 + 3500(P/A,8%,12) + 8000(P/F,8%,12)
= $-447

The decision is sensitive to the life of the investment.

18.10 At i = 5%, find the AW value for n from 1 to 15.

$$AW = -8000(A/P,5\%,n) - 500 - G(G/A,5\%,n)$$

For spreadsheet analysis, use the PMT functions to obtain the AW for each n value for each G amount. The table below includes the analysis for G = $60, $100 and $140. As an example, the cell entries for G = $-60 are:

For n = 1 year
A5:   1
B5: $–500
C5:   = -PMT(5%,1,NPV(5%,B5:B5)+B1)

For n = 12 years
A16: 12
B16: B15 – 60
C16: = -PMT(5%,12,NPV(5%,B5:B16)+B1)

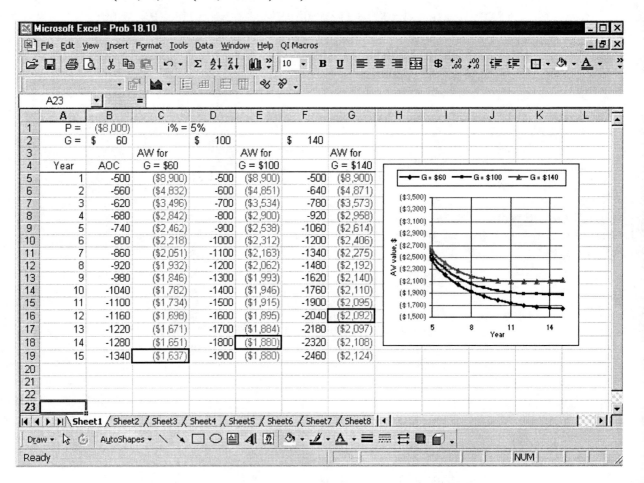

Results from columns C, E, and G are:

G	n*	AW
$60	15	$1637
100	14	1880
140	12	2092

The AW curves are quite flat; there are only a few dollars difference for the various n values around the n* value for each gradient value. The plot clearly shows this.

18.11 The PW relations are:
$$PW_A = -P_A + (R_A - AOC_A)(P/A,20\%,5) + 50,000(P/F,20\%,5)$$
$$PW_B = -P_B + (R_B - AOC_B)(P/A,20\%,5) + 37,000(P/F,20\%,5)$$

Tabular results are presented below:

(a) First cost

	A		B	
Variation	Value	PW$_A$	Value	PW$_B$
−50%	$250,000	$−5610	$187,500	$−23,100
0.00	500,000	−255,610	375,000	−210,600
100%	1,000,000	−755,610	750,000	−585,600

(b) AOC

	A		B	
Variation	Value	PW$_A$	Value	PW$_B$
−50%	$37,500	$−143,463	$40,000	$−90,976
0.00	75,000	−255,610	80,000	−210,600
100%	150,000	−479,905	160,000	−449,848

(c) Revenue

	A		B	
Variation	Value	PW$_A$	Value	PW$_B$
−50%	$75,000	$−479,905	$65,000	$−404,989
0.00	150,000	−255,610	130,000	−210,600
100%	300,000	+192,980	260,000	+178,178

18.12 (a) Purchase price

Variation	Value, P	ROR
−25%	$18,750	10.53%
0.00	25,000	1.91%
+25%	31,250	−4.47%

(IRR function)

$0 = P - 5500(P/F,i,1) - 1500(P/F,i,2) - 1300(P/F,i,3) + 35,000(P/F,i,3)$

Year	0	1	2	3
Cash flow, $	−P	−5500	−1500	33,700

(b) Selling price

Variation	Salvage, S	ROR
−25%	$26,250	−8.74%
0.00	35,000	1.91%
+25%	43,750	10.83%

(IRR function)

$0 = -25,000 - 5500(P/F,i,1) - 1500(P/F,i,2) - 1300(P/F,i,3) + S(P/F,i,3)$

Year	0	1	2	3
Cash flow, $	−25,000	−5500	−1500	S−1300

18.13 (a) First cost

$AW = -P(A/P,18\%,10) + 10,000(A/F,18\%,10) + 24,000$
$\quad\quad = -P(0.22251) + 24,425$

(b) AOC

$AW = -80,000(A/P,18\%,10) + 10,000(A/F,18\%,10) - AOC + 39,000$
$\quad\quad = -AOC + 21,624$

(c) Revenue

$AW = -80,000(A/P,18\%,10) + 10,000(A/F,18\%,10) - 15,000 + Revenue$
$\quad\quad = -32,376 + Revenue$

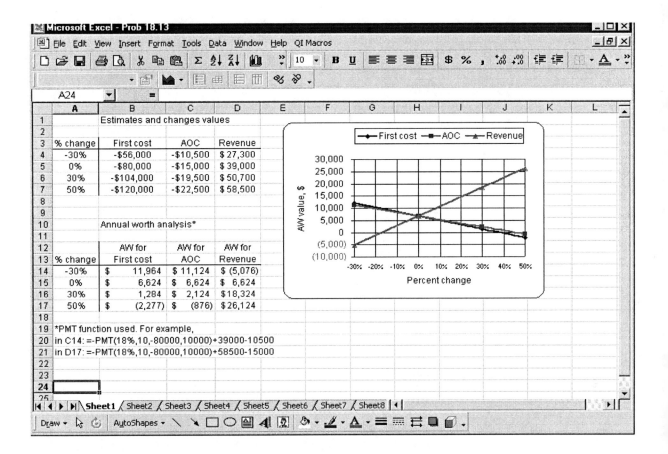

18.14 PW calculates the amount you should be willing to pay now. Plot PW versus ± 30% changes in (a), (b) and (c) on one graph.

(a) Face value, P

$$PW = P(P/F,4\%,20) + 450(P/A,4\%,20)$$
$$= P(0.4564) + 6116$$

(b) Dividend rate, b

$$PW = 10,000(P/F,4\%,20) + (10,000/2)(b)(P/A,4\%,20)$$
$$= 10,000(0.4564) + b(5000)(13.5903)$$
$$= 4564 + b(67,952)$$

(c) Nominal rate, r

$$PW = 10,000(P/F,r,20) + 450(P/A,r,20)$$

Chapter 18

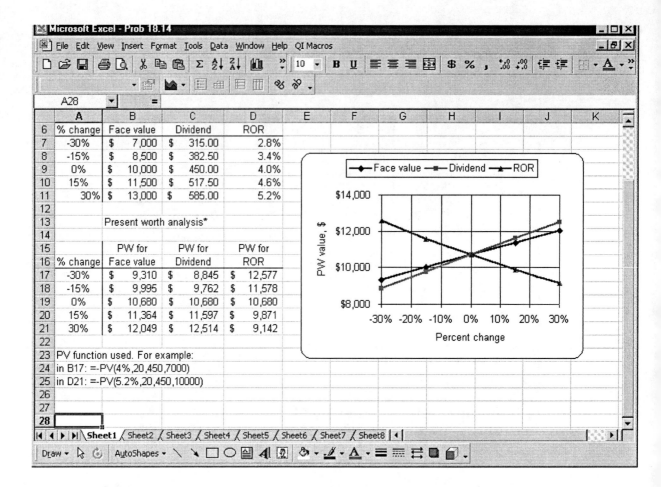

18.15 (a) 50 days

Plan 1 - Purchase

Opt: $0.40 per ton (AOC = $2000)
$$AW = -6000(A/P, 12\%, 5) - 0.40(100)(50)$$
$$= -6000(0.27741) - 2000$$
$$= \$-3664$$

ML: $0.50 per ton (AOC = $2500)
$$AW = -6000(A/P, 12\%, 5) - 0.50(100)(50)$$
$$= -6000(0.27741) - 2,500$$
$$= \$-4164$$

Pess: $0.75 per ton (AOC = $3750)
$$AW = -6000(A/P, 12\%, 5) - 0.75(100)(50)$$
$$= -6000(0.27741) - 3750$$
$$= \$-5414$$

Plan 2 - Lease

<u>Opt: $1800 lease</u>
AW = –1800 – 50(8)(5.00) = $–3800

<u>ML: $2500 lease</u>
AW = –2500 – 50(8)(5.00) = $–4500

<u>Pess: $3200 lease</u>
AW = –3200 – 50(8)(5.00) = $–5200

Plan 1 is better for the most likely estimates ($0.50 and $2500).

(b) 100 days   ### Plan 1 - Purchase

<u>Opt: $0.40 per ton (AOC = $4000)</u>
AW = –6000(A/P,12%,5) – 0.40(100)(100)
   = –6000(0.27741) – 4000
   = $–5664

<u>ML: $0.50 per ton (AOC = $5000)</u>
AW = –6000(A/P,12%,5) – 0.50(100)(100)
   = –6000(0.27741) – 5000
   = $–6664

<u>0.75 per ton (AOC = $7500)</u>

AW = –6000(A/P,12%,5) – 0.75(100)(100)
   = –6000(0.27741) – 7500
   = $–9164

Plan 2 - Lease

<u>Opt: $1800 lease</u>
AW = –1800 – 100(8)(5.00) = $–5800

<u>ML: $2,500 lease</u>
AW = –2500 – 100(8)(5.00) = $–6500

<u>Pess: $3,200 lease</u>
AW = –3200 – 100(8)(5.00) = $–7200

Plan 2 is better on the basis of the most likely estimates.

18.16 Water/wastewater cost = (0.12 + 0.04) per 1000 liters
= 0.16 per 1000 liters

Spray Method

Pessimistic - 100 liters

Water required = 10,000,000(100) = 1.0 billion

$AW = -(0.16/1000)(1.00 \times 10^9) = \$-160,000$

Most Likely - 80 liters

Water required = 10,000,000(80) = 800 million

$AW = -(0.16/1000)(800,000,000) = \$-128,000$

Optimistic - 40 liters

Water required = 10,000,000(40) = 400 million

$AW = -(0.16/1000)(400,000,000) = \$-64,000$

Immersion Method

$AW = -10,000,000(40)(0.16/1000) - 2000(A/P,15\%,10) - 100$
$= -64,000 - 2000(0.19925) - 100$
$= \$-64,499$

The immersion method is cheaper than the spray method, unless the optimistic estimate of 40 L is actually correct.

18.17 (a)   MARR = 8% (Pessimistic)

$PW_M = -100,000 + 15,000(P/A,8\%,20)$
$= -100,000 + 15,000(9.8181)$
$= \$47,272$

$PW_Q = -110,000 + 19,000(P/A,8\%,20)$
$= -110,000 + 19,000(9.8181)$
$= \$76,544$

MARR = 10% (Most Likely)

$PW_M = -100,000 + 15,000(P/A,10\%,20)$
$= -100,000 + 15,000(8.5136)$
$= \$27,704$

$PW_Q = -110,000 + 19,000(P/A,10\%,20)$
$= -110,000 + 19,000(8.5136)$
$= \$51,758$

<u>MARR = 15% (Optimistic)</u>

$PW_M = -100,000 + 15,000(P/A,15\%,20)$
$= -100,000 + 15,000(6.2593)$
$= \$-6111$

$PW_Q = -110,000 + 19,000(P/A,15\%,20)$
$= -110,000 + 19,000(6.2593)$
$= \$8927$

(b) <u>n = 16; Expanding economy (Optimistic)</u>

$n = 20(0.80) = 16$ years

$PW_M = -100,000 + 15,000(P/A,10\%,16)$
$= -100,000 + 15,000(7.8237)$
$= \$17,356$

$PW_Q = -110,000 + 19,000(P/A,10\%,16)$
$= -110,000 + 19,000(7.8237)$
$= \$38,650$

<u>n = 20; Expected economy (Most likely)</u>

$PW_M = \$27,704$ (From part (a))

$PW_Q = \$51,758$ (From part (a))

<u>n = 22; Receding economy (Pessimistic)</u>

$n = 20(1.10) = 22$ years

$PW_M = -100,000 + 15,000(P/A,10\%,22)$
$= -100,000 + 15,000(8.7715)$
$= \$31,573$

$PW_Q = -110,000 + 19,000(P/A,10\%,22)$
$= -110,000 + 19,000(8.7715)$
$= \$56,659$

(c) Plot the PW values for each value of MARR and life. Plan M always has a lower PW value, so it is not accepted and plan Q is.

18.18 $E(\text{flow}_N) = 0.15(100) + 0.75(200) + 0.10(300)$
               $= 195$ barrels/day

         $E(\text{flow}_E) = 0.35(100) + 0.15(200) + 0.45(300) + 0.05(400)$
               $= 220$ barrels/day

18.19 (a) $E(\text{time}) = (1/4)(10 + 20 + 30 + 70) = 32.5$ seconds
      (b) $E(\text{time}) = (1/3)(10 + 20 + 30) = 20$ seconds

Yes, the 70 second estimate does increase the mean significantly.

18.20

n	1	2	3	4
Y	3	9	27	81

       $E(Y) = 3(0.4) + 9(0.3) + 27(0.233) + 81(0.067)$
            $= 15.618$

18.21 Solve for the low AOC from E(AOC)

       $E(\text{AOC}) = 4575 = 2800(0.25) + (\text{high AOC})(0.75)$

       High AOC = $5167

18.22 $E(i) = 1/20[(-8)(1) + (-5)(1) + 0(5) + ... + 15(3)]$
          $= 103/20 = 5.15\%$

18.23 $E(\text{AW}) = 0.15(300{,}000 - 25{,}000) + 0.7(50{,}000)$
            $= \$76{,}250$

18.24 (a)   The subscripts identify the series by probability.

         $PW_{0.5} = -5000 + 1000(P/A, 20\%, 3)$
                 $= -5000 + 1000(2.1065)$
                 $= \$-2894$

         $PW_{0.2} = -6000 + 500(P/F, 20\%, 1) + 1500(P/F, 20\%, 2) + 2000(P/F, 20\%, 3)$
                 $= -6000 + 500(0.8333) + 1500(0.6944) + 2000(0.5787)$
                 $= \$-3384$

$$PW_{0.3} = -4000 + 3000(P/F,20\%,1) + 1200(P/F,20\%,2) - 800(P/F,20\%,3)$$
$$= -4000 + 3000(0.8333) + 1200(0.6944) - 800(0.5787)$$
$$= \$-1130$$

$$E(PW) = (PW_{0.5})(0.5) + (PW_{0.2})(0.2) + (PW_{0.3})(0.3)$$
$$= -2894(0.5) - 3384(0.2) - 1130(0.3)$$
$$= \$-2463$$

(b) $\quad E(AW) = E(PW)(A/P,20\%,3)$
$$= -2463(0.47473)$$
$$= \$-1169$$

18.25 Determine E(AW) after calculating E(revenue).

$$E(\text{revenue}) = 3[(\text{no. days})(\text{no. climbers})(\text{income/climbers})](\text{probability})$$
$$= [(120)(350)(5)](0.3) + [(120)(350)(5) + 30(100)(5)](0.5)$$
$$\quad + [(120)(350)(5) + (45)(100)(5)](0.2)$$
$$= 63{,}000 + 112{,}500 + 46{,}500$$
$$= \$222{,}000$$

$$E(AW) = -375{,}000(A/P,12\%,10) - 25{,}000[(P/F,12\%,4) + (P/F,12\%,8)]$$
$$\quad (A/P,12\%,10) - 56{,}000 + 222{,}000$$
$$= -375{,}000(0.17698) - 25{,}000[(0.6355) + (0.4039)](0.17698) + 166{,}000$$
$$= \$95{,}034$$

The mock mountain should be constructed.

18.26 Determine E(PW) after calculating the PW of E(revenue).

$$E(\text{revenue}) = P(\text{slump})(\text{revenue over 3-year periods})$$

$$PW(E(\text{revenue})) = PW\,[P(\text{slump})(\text{revenue 1}^{st}\text{ 3 years})$$
$$\quad + P(\text{slump})(\text{revenue 2}^{nd}\text{ 3 years})$$
$$\quad + P(\text{expansion})(\text{revenue 1}^{st}\text{ 3 years})$$
$$\quad + P(\text{expansion})(\text{revenue 2}^{nd}\text{ 3 years})]$$

$$= 0.5[20{,}000(P/A,8\%,3)] + 0.2[20{,}000(P/A,8\%,3)(P/F,8\%,3)] + 0.5[35{,}000(P/A,8\%,3)]$$
$$\quad + 0.8[35{,}000(P/A,8\%,3)(P/F,8\%,3)]$$

$$= 0.5[51{,}542] + 0.2[40{,}914] + 0.5[90{,}198] + 0.8[71{,}600]$$

$$= \$136{,}333$$

$$E(PW) = -200{,}000 + 200{,}000(0.12)\,(P/F,8\%,6) + PW(E(revenue))$$
$$= -200{,}000 + 15{,}125 + 136{,}333$$
$$= \$-48{,}542$$

No, less than an 8% return is expected.

18.27

Certificate of Deposit

Rate of return = 6.35% (from problem statement)

Stocks

<u>Stock 1</u>: $-5000 + 250(P/A,i\%,4) + 6800(P/F,i\%,5) = 0$
  is the i^* relation.
  $i^* = 10.07\%$ (RATE function)

<u>Stock 2</u>: $-5000 + 600(P/A,i\%,4) + 4000(P/F,i\%,5) = 0$
  $i^* = 6.36\%$ (RATE function)

$E(i) = 10.07(0.5) + 6.36(0.5) = 8.22\%$

Real Estate

<u>Rate of return with Prob = 0.3</u>
  $-5{,}000 - 425(P/A,i\%,4) + 9500(P/F,i\%,5) = 0$
  $i^* = 8.22\%$

<u>Rate of return with Prob. 0.5</u>
  $-5000 + 7200(P/F,i\%,5) = 0$
  $(P/F,i\%,5) = 0.6944$
  $i^* = 7.57\%$

<u>Rate of return with Prob. 0.2</u>
  $-5000 + 500(P/A,i\%,4) + 100(P/G,i\%,4) + 5200(P/F,i\%,5) = 0$
  $i^* = 11.34\%$

$E(i) = 8.22(0.3) + 7.57(0.5) + 11.34(0.2)$
   $= 8.52\%$

Invest in real estate for the highest E(rate of return) of 8.52%.

18.28 (a) Calculate fraction in equity times i on equity from graph.

$$E(i) = 0.3(i \text{ on } 20\text{-}80) + 0.5(i \text{ on } 50\text{-}50) + 0.2(i \text{ on } 80\text{-}20)$$
$$= 0.3(7\%) + 0.5(9\%) + 0.2(11.5)$$
$$= 8.9\%$$

(b) (Fraction of pool)($1 million)(fraction of D-E in equity)

$$\text{Amount} = 0.3(\$1 \text{ mil})(0.8) + 0.5(\$1 \text{ mil})(0.5) + 0.2(\$1 \text{ mil})(0.2)$$
$$= 0.3 (800{,}000) + 0.5(500{,}000) + 0.2(200{,}000)$$
$$= 240{,}000 + 250{,}000 + 40{,}000$$
$$= \$530{,}000$$

The FW is calculated using the correct i rate for each equity amount.

$$\text{FW} = 240{,}000(F/P,7\%,10) + 250{,}000(F/P,9\%,10) + 40{,}000(F/P,11.5\%,10)$$
$$= 240{,}000(1.9672) + 250{,}000(2.3674) + 40{,}000(2.9699)$$
$$= \$1{,}182{,}755$$

(c) Use Eq. [14.9] to determine the real i. The graph rates are actually i_f values.

$$\text{at } i_f = 7\%: i = (i_f - f)/(1 + f)$$
$$= (0.07 - 0.045)/(1 + 0.045)$$
$$= 0.0239 \quad\quad (2.39\%)$$

$$\text{at } i_f = 9\%: i = (0.09 - 0.045)/1.045 = 0.043 \quad (4.3\%)$$

$$\text{at } i_f = 11.5\%: i = (0.115 - 0.045)/1.045 = 0.067 \quad (6.7\%)$$

This is case 2 in Sec. 14.3. Use Eq. [14.8].

$$\text{FW} = 1{,}182{,}755/(1.045)^{10}$$
$$= 1{,}182{,}755/1.55297$$
$$= \$761{,}608$$

Alternatively, find PW at the real i for each equity amount.

$$\text{FW} = 240{,}000(F/P,2.39\%,10) + 250{,}000(F/P,4.3\%,10)$$
$$\quad\quad + 40{,}000(F/P,6.7\%,10)$$
$$= 240{,}000(1.26641) + 250{,}000(1.5238) + 40{,}000(1.9127)$$
$$= \$303{,}939 + 380{,}876 + 76{,}508$$
$$+ \$761{,}323 \quad\quad \text{(Rounding of i makes the difference)}$$

18.29 AW = annual loan payment + (damage) x P(rainfall amount or greater)

The subscript on AW indicates the rainfall amount.

$AW_{2.0}$ = –200,000(A/P,6%,10) + (–50,000)(0.3)
= –200,000(0.13587) –50,000(0.3)
= $–42,174

$AW_{2.25}$ = –225,000(A/P,6%,10) + (–50,000)(0.1)
= –300,000(0.13587) –50,000(0.1)
= $–35,571

$AW_{2.5}$ = –300,000(A/P,6%,10) + (–50,000)(0.05)
= –350,000(0.13587) –50,000(0.05)
= $–43,261

$AW_{3.0}$ = –400,000(A/P,6%,10) + (–50,000)(0.01)
= –400,000(0.13587) –50,000(0.01)
= $–54,848

$AW_{3.25}$ = –450,000(A/P,6%,10) + (–50,000)(0.005)
= –450,000(0.13587) –50,000(0.005)
= $–61,392

Build a wall to protect against a rainfall of 2.25 inches with an expected AW of $–35,571.

18.30 Compute the expected value for each outcome and select the minimum for D3.

Top node:
0.2(55) + 0.35(–30) + 0.45(10) = 5.0

Bottom node:
0.4(–17) + 0.6(0) = –6.8

Indicate 5.0 and –6.8 in ovals and select the top branch with E(value) = 5.0.

18.31 Maximize the value at each decision node.

D3: Top: E(value) = $30
     Bottom: E(value) = 0.4(100) + 0.6(–50) = $10

Select top at D3 for $30

D1: Top: $0.9(D3\ value) + 0.1(final\ value)$
           $0.9(30) + 0.1(500) = \$77$
   Value at D1 = 77-50 = \$27
Bottom: $90 - 80 = \$10$

Select top at D1 for \$27

D2: Top: $E(value) = 0.3(150 - 30) + 0.4(75) = \$66$
Middle: $E(value) = 0.5(200 - 100) = \$50$
Bottom: $E(value) = \$50$

At D2, value = E(value) – investment
Top: $66 - 25 = \$41$ (maximum)
Middle: $50 - 30 = \$20$
Bottom: $50 - 20 = \$30$

Select top at D2 for \$41

Conclusion: Select D2 path and choose top branch (\$25 investment)

18.32 Calculate the E(PW) in year 3 and select the largest expected value. In \$1000 terms:

$$E(PW\ of\ D4,x) = -200 + 0.7[50(P/A,15\%,3)] + 0.3[40(P/F,15\%,1) \\ + 30(P/F,15\%,2) + 20(P/F,15\%,3)]$$

$$= -98.903 \qquad (\$-98{,}903)$$

$$E(PW\ of\ D4,y) = -75 + 0.45[30(P/A,15\%,3) + 10(P/G,15\%,3)] \\ + 0.55[30(P/A,15\%,3)]$$

$$= 2.816 \qquad (\$2816)$$

$$E(PW\ of\ D4,z) = -350 + 0.7[190(P/A,15\%,3) - 20(P/G,15\%,3)] \\ + 0.3[-30(P/A,15\%,3)]$$

$$= -95.880 \qquad (\$-95{,}880)$$

Select decision branch y; it has the largest E(PW).

18.33 Select the minimum E(cost) alternative. (All dollar values are times $1000).

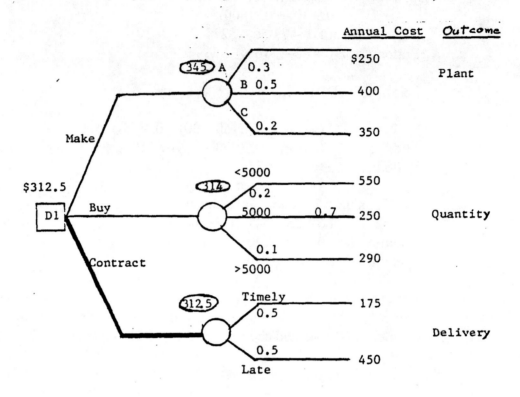

$$E(\text{cost of plant}) = 0.3(250) + 0.5(400) + 0.2(350)$$
$$= \$345 \qquad (\$345{,}000)$$

$$E(\text{cost of quantity}) = 0.2(550) + 0.7(250) + 0.1(290)$$
$$= \$314 \qquad (\$314{,}000)$$

$$E(\text{cost of delivery}) = 0.5(175 + 450)$$
$$= \$312.5 \qquad (\$312{,}500)$$

Select the contract alternative since the E(cost of delivery) is the lowest.

18.34 (a) Construct the decision tree.

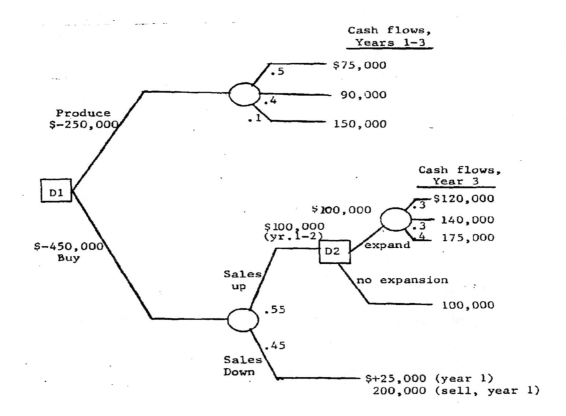

(b) At D2 compute PW of cash flows and E(PW) using probability values.

Expansion option

(PW for D2, $120,000) = –100,000 + 120,000(P/F,15%,1)
 = $4352
(PW for D2, $140,000) = –100,000 + 140,000(P/F,15%,1)
 = $21,744
(PW for D2, $175,000) = $52,180

E(PW) = 0.3(4352 + 21,744) + 0.4(52,180) = $28,700

No expansion option

(PW for D2, $100,000 = $100,000(P/F,15%,1) = $86,960
 E(PW) = $86,960

Conclusion at D2: Select no expansion option

(c) Complete foldback to D1 considering 3 year cash flow estimates.

Produce option, D1

$$E(PW \text{ of cash flows}) = [0.5(75,000) + 0.4(90,000) + 0.1(150,000)](P/A,15\%,3)$$
$$= \$202,063$$

$$E(PW \text{ for produce}) = \text{cost} + E(PW \text{ of cash flows})$$
$$= -250,000 + 202,063$$
$$= \$-47,937$$

Buy option, D1

At D2, $E(PW) = \$86,960$

$$E(PW \text{ for buy}) = \text{cost} + E(PW \text{ of sales cash flows})$$
$$= -450,000 + 0.55(PW \text{ sales up}) + 0.45(PW \text{ sales down})$$

PW Sales up $= 100,000(P/A,15\%,2) + 86,960(P/F,15\%,2)$
$= \$228,320$

PW sales down $= (25,000 + 200,000)(P/F,15\%,1)$
$= \$195,660$

$$E(PW \text{ for buy}) = -450,000 + 0.55(228,320) + 0.45(195,660)$$
$$= \$-236,377$$

Conclusion: E(PW for produce) is larger than E(PW for buy); select produce option.

Note: The returns are both less than 15%, but the return is larger for produce option than buy.

(d) The return would increase on the initial investment, but would increase faster for the produce option.

Extended Exercise Solution

1. Relations are developed here for hand solution.

$$\underline{MARR = 8\%}$$

PW_A = $-10,000 + 1000(P/F,8\%,40) - 500(P/A,8\%,40)$
       = $-10,000 + 1000(0.0460) - 500(11.9246)$
       = $\$-15,916$

PW_B = $-30,000 + 5000(P/F,8\%,40) - 100(P/A,8\%,40) - 5000$
        $- 200(P/F,8\%,20) - 5000(P/F,8\%,20) - 200(P/F,8\%,40) - 200(P/A,8\%,40)$
       = $-35,000 + 4800(P/F,8\%,40) - 300(P/A,8\%,40) - 5200(P/F,8\%,20)$
       = $-35,000 + 4800(0.0460) - 300(11.9246) - 5200(0.2145)$
       = $\$-39,472$

$$\underline{MARR = 10\%}$$

PW_A = $-10,000 + 1000(P/F,10\%,40) - 500(P/A,10\%,40)$
       = $-10,000 + 1000(0.0221) - 500(9.7791)$
       = $\$-14,867$

PW_B = $-30,000 + 5000(P/F,10\%,40) - 100(P/A,10\%,40) - 5000$
        $- 200(P/F,10\%,20) - 5000(P/F,10\%,20) - 200(P/F,10\%,40)$
        $- 200(P/A,10\%,40)$
       = $-35,000 + 4800(P/F,10\%,40) - 300(P/A,10\%,40) - 5200(P/F,10\%,20)$
       = $-35,000 + 4800(0.0221) - 300(9.7791) - 5200(0.1486)$
       = $\$-38,600$

$$\underline{MARR = 15\%}$$

PW_A = $-10,000 + 1000(P/F,15\%,40) - 500(P/A,15\%,40)$
       = $-10,000 + 1000(0.0037) - 500(6.6418)$
       = $\$-13,317$

PW_B = $-30,000 + 5000(P/F,15\%,40) - 100(P/A,15\%,40) - 5000$
        $-200(P/F,15\%,20) - 5000(P/F,15\%,20) - 200(P/F,15\%,40)$
        $-200(P/A,15\%,40)$
       = $-35,000 + 4800(P/F,15\%,40) - 300(P/A,15\%,40) - 5200(P/F,15\%,20)$
       = $-35,000 + 4800(0.0037) - 300(6.6418) - 5200(0.0611)$
       = $\$-37,293$

Not very sensitive.

2.

Expanding economy

$n_A = 40(0.80) = 32$ years
$n_1 = 40(0.80) = 32$ years
$n_2 = 20(0.80) = 16$ years

$PW_A = -10,000 + 1000(P/F,10\%,32) - 500(P/A,10\%,32)$
$\quad\quad = -10,000 + 1,000(0.0474) - 500(9.5264)$
$\quad\quad = \$-14,716$

$PW_B = -30,000 + 5000(P/F,10\%,32) - 100(P/A,10\%,32) - 5000$
$\quad\quad -200(P/F,10\%,16) - 5000(P/F,10\%,16) - 200(P/F,10\%,32)$
$\quad\quad -200(P/A,10\%,32)$
$\quad\quad = -35,000 + 4800(P/F,10\%,32) - 300(P/A,10\%,32) - 5200(P/F,10\%,16)$
$\quad\quad = -35,000 + 4800(0.0474) - 300(9.5264) - 5200(0.2176)$
$\quad\quad = \$-38,762$

Expected economy

$PW_A = \$-14,876$ (from #1)

$PW_B = \$-38,600$ (from #1)

Receding economy

$n_A = 40(1.10) = 44$ years
$n_1 = 40(1.10) = 44$ years
$n_2 = 20(1.10) = 22$ years

$PW_A = -10,000 + 1000(P/F,10\%,44) - 500(P/A,10\%,44)$
$\quad\quad = -10,000 + 1000(0.0154) - 500(9.8461)$
$\quad\quad = \$-14,908$

$PW_B = -30,000 + 5000(P/F,10\%,44) - 100(P/A,10\%,44) - 5000$
$\quad\quad - 200(P/F,10\%,22) - 5000(P/F,10\%,22) - 200(P/F,10\%,44)$
$\quad\quad - 200(P/F,10\%,44)$
$\quad\quad = -35,000 + 4800(P/F,10\%,44) - 300(P/A,10\%,44) - 5200(P/F,10\%,22)$
$\quad\quad = -35,000 + 4800(0.0154) - 300(9.8461) - 5200(0.1228)$
$\quad\quad = \$-38,519$

Not very sensitive.

3. In all cases, plan A has the best PW.

4. Use SOLVER to find the breakeven values of P_A for the three MARR values of 8%, 10%, and 15% per year.

    For MARR = 8%, the SOLVER screen is below.

    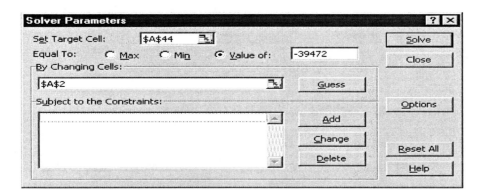

    Breakeven values are:

MARR	Breakeven P_A
8%	$–33,556
10	–33,734
15	–33,975

    The P_A breakeven value is not sensitive, but all three outcomes are over 3X the $10,000 estimated first cost for plan A.

Case Study Solution

1. Let x = weighting per factor

   Since there are 6 factors and one (environmental considerations) is to have a weighting that is double the others, its weighting is 2x. Thus,

   $$2x + x + x + x + x + x = 100$$
   $$7x = 100$$
   $$x = 14.3\%$$

   Therefore, the environmental weighting is 2(14.3), or 28.6%

2.

Alt ID	Ability to Supply Area	Relative Cost	Engineering Feasibility	Institutional Issues	Environmental Considerations	Lead-Time Requirement	Total
1A	5(0.2)	4(0.2)	3(0.15)	4(0.15)	5(0.15)	3(0.15)	4.1
3	5(0.2)	4(0.2)	4(0.15)	3(0.15)	4(0.15)	3(0.15)	3.9
4	4(0.2)	4(0.2)	3(0.15)	3(0.15)	4(0.15)	3(0.15)	3.6
8	1(0.2)	2(0.2)	1(0.15)	1(0.15)	3(0.15)	4(0.15)	2.0
12	5(0.2)	5(0.2)	4(0.15)	1(0.15)	3(0.15)	1(0.15)	3.4

   Therefore, the top three are the same as before: 1A, 3, and 4

3. For alternative 4 to be as economically attractive as alternative 3, its total annual cost would have to be the same as that of alternative 3, which is $3,881,879. Thus, if P_4 is the capital investment,

   $$3,881,879 = P_4(A/P, 8\%, 20) + 1,063,449$$
   $$3,881,879 = P_4(0.10185) + 1,063,449$$
   $$P_4 = \$27,672,361$$

   Decrease = 29,000,000 − 27,672,361
   = $1,327,639 or 4.58%

4. Household cost at 100% = 3,952,959(1/12)(1/4980)(1/1)
   = $66.15

   Decrease = 69.63 − 66.15
   = $3.48 or 5%

5. (a) Sensitivity analysis of M&O and number of households.

Alternative	Estimate	M&O, $/year	Number of households	Total annual cost, $/year	Household cost, $/month
1A	Pessimistic	1,071,023	4980	3,963,563	69.82
	Most likely	1,060,419	5080	3,952,959	68.25
	Optimistic	1,049,815	5230	3,942,355	66.12
3	Pessimistic	910,475	4980	3,925,235	69.40
	Most likely	867,119	5080	3,881,879	67.03
	Optimistic	867,119	5230	3,881,879	**65.10**
4	Pessimistic	1,084,718	4980	4,038,368	71.13
	Most likely	1,063,449	5080	4,017,099	69.37
	Optimistic	957,104	5230	3,910,754	65.59

Conclusion: Alternative 3 – optimistic is the best.

(b) Let x be the number of households. Set alternative 4 – optimistic cost equal to $65.10.

$$(3,910,754)/12(0.95)(x) = \$65.10$$
$$x = 5270$$

This is an increase of only 40 households.

Chapter 19 – More on Variation and Decision Making Under Risk
Solutions to end of chapter exercises

Problems

19.1 (a) Continuous (assumed) and uncertain – no chance statements made.
(b) Discrete and risk – plot units vs. chance as a continuous straight line between 50 and 55 units.
(c) 2 variables: first is discrete and certain at $400; second is continuous for $\geq$ $400, but uncertain (at this point). More data needed to assign any probabilities.
(d) Discrete variable with risk; rain at 20%, snow at 30%, other at 50%.

19.2 Needed or assumed information to be able to calculate an expected value:
1. Treat output as discrete or continuous variable.
2. If discrete, center points on cells, e.g., 800, 1500, and 2200 units per week.
3. Probability estimates for < 1000 and /or > 2000 units per week.

19.3 (a) N is discrete since only specific values are mentioned; i is continuous from 0 to 12.
(b) The P(N), F(N), P(i) and F(i) are calculated below.

N	0	1	2	3	4
P(N)	.12	.56	.26	.03	.03
F(N)	.12	.68	.94	.97	1.00

i	0-2	2-4	4-6	6-8	8-10	10-12
P(i)	.22	.10	.12	.42	.08	.06
F(i)	.22	.32	.44	.86	.94	1.00

(c) $P(N = 1 \text{ or } 2) = P(N = 1) + P(N = 2)$
$= 0.56 + 0.26 = 0.82$
or
$F(N \leq 2) - F(N \leq 0) = 0.94 - 0.12 = 0.82$

$P(N \geq 3) = P(N = 3) + P(N \geq 4) = 0.06$

(d) $P(7\% \leq i \leq 11\%) = P(6.01 \leq i \leq 12)$
$= 0.42 + 0.08 + 0.06 = 0.56$
or
$F(i \leq 12\%) - F(i \leq 6\%) = 1.00 - 0.44$
$= 0.56$

19.4 (a)

$	0	2	5	10	100
F($)	.91	.955	.98	.993	1.000

The variable $ is discrete, so plot $ versus F($).

(b) $E(\$) = \Sigma \$ P(\$) = 0.91(0) + ... + 0.007(100)$
$= 0 + 0.09 + 0.125 + 0.13 + 0.7$
$= \$1.045$

(c) $2.000 - 1.045 = 0.955$
Long-term income is 95.5¢ per ticket

19.5 (a) $P(N) = (0.5)^N \qquad N = 1,2,3,...$

N	1	2	3	4	5	etc.
P(N)	0.5	0.25	0.125	0.0625	0.03125	
F(N)	0.5	0.75	0.875	0.9375	0.96875	

Plot P(N) and F(N); N is discrete.

P(L) is triangular like the distribution in Figure 19-5 with the mode at 5.

$$f(mode) = f(M) = \frac{2}{5-2} = \frac{2}{3}$$

$$F(mode) = F(M) = \frac{5-2}{5-2} = 1$$

(b) $P(N = 1, 2 \text{ or } 3) = F(N \leq 3) = 0.875$

19.6 First cost, P
P_P = first cost to purchase
P_L = first cost to lease

Use the uniform distribution relations in Equation [19.3] and plot.

$f(P_P) = 1/(25{,}000 - 20{,}000) = 0.0002$

$f(P_L) = 1/(2000 - 1800) = 0.005$

Salvage value, S

S_P is triangular with mode at $2500.

The $f(S_P)$ is symmetric around $2500.

$f(M) = f(2500) = 2/(1000) = 0.002$ is the probability at $2500.

There is no S_L distribution

AOC

$f(AOC_P) = 1/(9000-5000) = 0.00025$

$f(AOC_L)$ is triangular with:
$f(7000) = 2/(9000-5000) = 0.0005$

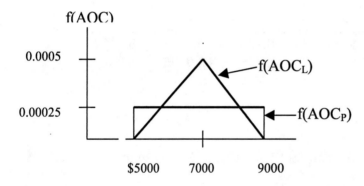

Life, L

$f(L_P)$ is triangular with mode at 6.

$f(6) = 2/(8-4) = 0.5$

The value L_L is certain at 2 years.

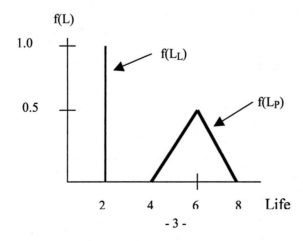

Chapter 19

19.7 (a) Determine several values of D_M and D_Y and plot.

D_M or D_Y	$f(D_M)$	$f(D_Y)$
0.0	3.00	0.0
0.2	1.92	0.4
0.4	1.08	0.8
0.6	0.48	1.2
0.8	0.12	1.6
1.0	0.00	2.0

$f(D_M)$ is a decreasing power curve and $f(D_Y)$ is linear.

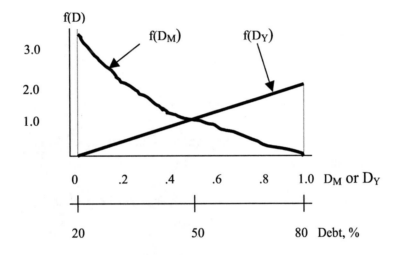

(b) Probability is larger that M (mature) companies have a lower debt percentage and that Y (young) companies have a higher debt percentage.

19.8 (a)

X_i	1	2	3	6	9	10
$F(X_i)$	0.2	0.4	0.6	0.7	0.9	1.0

(b) $P(6 \le X \le 10) = F(10) - F(3) = 1.0 - 0.6 = 0.4$

or

$P(X = 6, 9$ or $10) = 0.1 + 0.2 + 0.1 = 0.4$
$P(X = 4, 5$ or $6) = F(6) - F(3) = 0.7 - 0.6 = 0.1$

(c) $P(X = 7$ or $8) = F(8) - F(6) = 0.7 - 0.7 = 0.0$

No sample values in the 50 have X = 7 or 8. A larger sample is needed to observe all values of X.

19.9 Plot the $F(X_i)$ from Problem 19.8 (a), assign the RN values, use Table 19.2 to obtain 25 sample X values; calculate the sample $P(X_i)$ values and compare them to the stated probabilities in 19.8.

(Instructor note: Point out to students that it is not correct to develop the sample $F(X_i)$ from another sample where some discrete variable values are omitted).

19.10 (a)

X	0	.2	.4	.6	.8	1.0
F(X)	0	.04	.16	.36	.64	1.00

Take X and p values from the graph. Some samples are:

RN	X	p
18	.42	7.10%
59	.76	8.80
31	.57	7.85
29	.52	7.60

(b) Use the sample mean for the average p value. Our sample of 30 had p = 6.3375%; yours will vary depending on the RNs from Table 19.2.

19.11 Use the steps in Section 19.3. As an illustration, assume the probabilities that are assigned by a student are:

$$P(G = g) = \begin{bmatrix} 0.30 & G=A \\ 0.40 & G=B \\ 0.20 & G=C \\ 0.10 & G=D \\ 0.00 & G=F \\ 0.00 & G=I \end{bmatrix}$$

1 and 2. The F(G) and RN assignment are:

$$F(G = g) = \begin{bmatrix} 0.30 & G=A & 00\text{-}29 \\ 0.70 & G=B & 30\text{-}69 \\ 0.90 & G=C & 70\text{-}89 \\ 1.00 & G=D & 90\text{-}99 \\ 1.00 & G=F & -- \\ 1.00 & G=I & -- \end{bmatrix}$$

3 and 4. Develop a scheme for selecting the RNs from Table 19-2. Assume you want 25 values. For example, if $RN_1 = 39$, the value of G is B. Repeat for sample of 25 grades.

5. Count the number of grades A through D, calculate the probability of each as count/25, and plot the probability distribution for grades A through I. Compare these probabilities with $P(G=g)$ above.

19.12 (a) When RAND( ) was used for 100 values in column A of an Excel spreadsheet, the function AVERAGE(A1:A100) resulted in 0.50750658; very close to 0.5.

(b) Count the values in each cell to see how close it is to 10.

19.13 (a) Use Equations [19.9] and [19.12] or the spreadsheet functions AVERAGE and STDEV.

Cell, X_i	f_i	X_i^2	$f_i X_i$	$f_i X_i^2$
600	6	360,000	3,600	2,160,000
800	10	640,000	8,000	6,400,000
1000	9	1,000,000	9,000	9,000,000
1200	15	1,440,000	18,000	21,600,000
1400	28	1,960,000	39,200	54,880,000
1600	15	2,560,000	24,000	38,400,000
1800	7	3,240,000	12,600	22,680,000
2000	10	4,000,000	20,000	40,000,000
	100		134,400	195,120,000

AVERAGE: $\bar{0} = 134,400/100 = 1344.00$

STDEV: $s^2 = \left[\dfrac{195,120,000}{99} - \dfrac{100\,(1344)^2}{99}\right]^2$

$= (146,327.27)^2$
$= 382.53$

(b) $\bar{0} \pm 2s$ is $1344.00 \pm 2(382.53) = 578.94$ and 2109.06
All values are in the $\pm 2s$ range.

(c) Plot X versus f. Indicate $\bar{0}$ and the range $\bar{0} \pm 2s$ on it.

19.14 (a) Convert P(X) data to frequency values to determine s.

X	P(X)	XP(X)	f	X^2	fX^2
1	.2	.2	10	1	10
2	.2	.4	10	4	40
3	.2	.6	10	9	90
6	.1	.6	5	36	180
9	.2	1.8	10	81	810
10	.1	1.0	5	100	500
		4.6			1630

Sample average: $\bar{0} = 4.6$

Sample variance: $s^2 = \dfrac{1630}{49} - \dfrac{50}{49}(4.6)^2 = 11.67$

$s = 3.42$

(b) $\bar{0} \pm 1s$ is $4.6 \pm 3.42 = 1.18$ and 8.02
25 values, or 50%, are in this range.

$\bar{0} \pm 2s$ is $4.6 \pm 6.84 = -2.24$ and 11.44
All 50 values, or 100%, are in this range.

19.15 (a) Use Equations [19.15] and [19.16]. Substitute Y for D_Y.

$f(Y) = 2Y$

$E(Y) = \int_0^1 (Y)2Y\,dy$

$= \left[\dfrac{2Y^3}{3}\right]_0^1$

$= 2/3 - 0 = 2/3$

$Var(Y) = \int_0^1 (Y^2)2Y\,dy - [E(Y)]^2$

$= \left[\dfrac{2Y^4}{4}\right]_0^1 - (2/3)^2$

$= \dfrac{2}{4} - 0 - \dfrac{4}{9} = 1/18$

$\sigma = \sqrt{1/18} = 0.236$

(b) $E(Y) \pm 2\sigma$ is $0.667 \pm 0.472 = 0.195$ and 1.139

Take the integral from 0.195 to 1.0 only since the variable upper limit is 1.0.

$$P(0.195 \leq Y \leq 1.0) = \int_{0.195}^{1} 2Y\,dy$$

$$= Y^2 \Big|_{0.195}^{1}$$

$$= 1 - 0.038 = 0.962 \qquad (96.2\%)$$

19.16 (a) Use Equations [19.15] and [19.16]. Substitute M for D_M.

$$E(M) = \int_0^1 (M)\, 3(1-M)^2\,dm$$

$$= 3\int_0^1 (M - 2M^2 + M^3)\,dm$$

$$= 3\left[\frac{M^2}{2} - \frac{2M^3}{3} + \frac{M^4}{4}\right]_0^1$$

$$= \frac{3}{2} - 2 + \frac{3}{4} = \frac{6-8+3}{4} = \frac{1}{4} = 0.25$$

$$\mathrm{Var}(M) = \int_0^1 (M^2)\, 3(1-M)^2\,dm - [E(M)]^2$$

$$= 3\int_0^1 (M^2 - 2M^3 + M^4)\,dm - (1/4)^2$$

$$= 3\left[\frac{M^3}{3} - \frac{M^4}{2} + \frac{M^5}{5}\right]_0^1 - 1/16$$

$$= 1 - 3/2 + 3/5 - 1/16$$

$$= (80 - 120 + 48 - 5)/80$$

$$= 3/80 = 0.0375$$

$$\sigma = \sqrt{0.0375}$$

$$= 0.1936$$

(b) $E(M) \pm 2\sigma$ is $0.25 \pm 2(0.1936) = -0.1372$ and 0.6372

Use the relation defined in Problem 19.15 to take the integral from 0 to 0.6372.

$$P(0 \leq M \leq 0.6372) = \int_0^{0.6372} 3(1-M)^2\,dm$$

$$= 3 \int_0^{0.6372} (1 - 2M + M^2)\,dm$$

$$= 3\left[M - M^2 + 1/3\, M^3 \right]_0^{0.6372}$$

$$= 3\left[0.6372 - (0.6372)^2 + 1/3\,(0.6372)^3 \right]$$
$$= 0.952 \qquad (95.2\%)$$

19.17 Use Eq. [19.8] where $P(N) = (0.5)^N$

$E(N) = \Sigma NP(N) = 1(.5) + 2(.25) + 3(.125) + 4(0.625)$
  $+ 5(.03125) + 6(.015625) + 7(.0078125)$
  $+ 8(.003906) + 9(.001953) + 10(.0009766) + \ldots$
  $= 1.99+$

E(N) can be calculated for as many N values as you wish. The limit to the series $N(0.5)^N$ is 2.0, the correct answer.

19.18 $E(Y) = \Sigma YP(Y) = 3(1/3) + 7(1/4) + 10(1/3) + 12(1/12)$
  $= 1 + 1.75 + 3.333 + 1$
  $= 7.083$

Var$(Y) = \Sigma Y^2 P(Y) - [E(Y)]^2$
  $= 3^2(1/3) + 7^2(1/4) + 10^2(1/3) + 12^2(1/12) - (7.083)^2$
  $= 60.583 - 50.169$
  $= 10.414$

$\sigma = 3.227$

$E(X) \pm 1\sigma$ is $7.083 \pm 3.227 = 3.856$ and 10.31

19.19 Using a spreadsheet, the steps in Sec. 19.5 are applied.

1. CFAT given for years 0 through 6.
2. i varies between 6% and 10%.
   CFAT for years 7-10 varies between $1600 and $2400.
3. Uniform for both i and CFAT values.
4. Set up a spreadsheet. The example below has the following relations:

   Col A: =RAND ( )* 100 to generate random numbers from 0-100.
   Col B, cell B4: =INT((.04*A4+6) *100)/10000 converts the RN to i from 0.06 to 0.10. The % designation changes it to an interest rate between 6% and 10%.
   Col C: = RAND( )* 100
   Col D, cell D4: =INT (8*C4+1600) to convert to a CFAT between $1600 and $2400.
   Ten samples of i and CFAT for years 7-10 are shown below in columns B and D of the spreadsheet.

A	B	C	D	E	F	G
RN for i	i	RN for CFAT	CFAT, years 7-10	Year	Annual CFAT using D4 for CFAT and B4 for MARR	Annual CFAT using D5 for CFAT and B5 for MARR
97.0043	9.88%	24.4147	$ 1,795	0	($28,800)	($28,800)
0.58075	6.02%	24.6312	$ 1,797	1	$ 5,400	$ 5,400
42.9306	7.71%	22.558	$ 1,780	2	$ 5,400	$ 5,400
42.4314	7.69%	62.8228	$ 2,102	3	$ 5,400	$ 5,400
39.5707	7.58%	4.09544	$ 1,632	4	$ 5,400	$ 5,400
44.7825	7.79%	42.4287	$ 1,939	5	$ 5,400	$ 5,400
29.6074	7.18%	13.6669	$ 1,709	6	$ 5,400	$ 5,400
97.9149	9.91%	46.9506	$ 1,975	7	$ 1,795	$ 1,797
95.4244	9.81%	44.0617	$ 1,952	8	$ 1,795	$ 1,797
84.159	9.36%	51.482	$ 2,011	9	$ 1,795	$ 1,797
				10	$ 4,595	$ 4,597
			PW of CFAT		($866)	$3,680

5. Columns F and G give two of the CFAT sequences, for example only, using rows 4 and 5 random number generations. The entry for cells F11 through F13 is =D4 and cell F14 is =D4+2800, where S = $2800. The PW values are obtained using the spreadsheet NPV function. The value PW = $-866 results from the i value in B4 (i = 9.88%) and PW = $3680 results from applying the MARR in B5 (i = 6.02%).

6. Plot the PW values for as large a sample as desired. Or, following the logic of Figure 19-13, a spreadsheet relation can count the + and – PW values, with $\bar{0}$ and s calculated for the sample.

7. Conclusion:
    For certainty, accept the plan since PW = $2966 exceeds zero at an MARR or 7% per year.
    For risk, the result depends on the preponderance of positive PW values from the simulation, and the distribution of PW obtained in step 6.

19.20 Use the spreadsheet Random Number Generator (RNG) on the tools toolbar to generate CFAT values in column D from a normal distribution with Φ = $2040 and σ = $500. The RNG screen image is shown below. (This tool may not be available on all spreadsheets.)

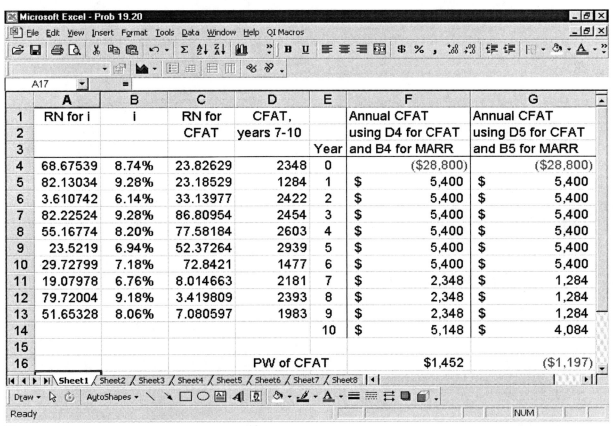

The spreadsheet above is the same as that in Problem 19.19, except that FAT values in column D for years 7 through 10 are generated using the RNG for the normal distribution described above. Te decision to accept the plan uses the same logic as that described in 19.19.

Extended Exercise Solution

This simulation is left to the student and the instructor. The same 7-step procedure from Section 19.5 applied in Problems 19.19 and 19.20 is used to set up the RNG for the cash flow values AOC and S, and the alternative life n for each alternative. The distributions given in the statement of the exercise are defined using the RNG.

For each of the 50-sample cash flow series, calculate the AW value for each alternative. To obtain a final answer of which alternative is the best to accept, it is recommended that the number of positive and negative AW values be counted as they are generated. Then the alternative with the most positive AW values indicates which one to accept. Of course, due to the RNG generation of AOC, S and n values, this decision may vary from one simulation run to the next.

Appendix B – Basics of Accounting Reports and Business Ratios
Solutions to end of appendix problems

B.1 (a)

Non-Stop
Balance Sheet
July 31, 20XX

Assets

Current
- Cash $17,000
- Accounts receivable   29,000
- Inventory   31,000
- Total current assets $77,000

Fixed
- Land $450,000
- Building   605,000
- Total fixed assets $1,055,000

Total assets $1,132,000

Liabilities

- Accounts payable $ 35,000
- Dividends payable   8,000
- Mortgage payable   450,000
- Bonds payable   110,000

Total liabilities $ 603,000

Net Worth

- Retained earnings $154,000
- Stock value   375,000
- Total net worth   529,000

Total liabilities and net worth $1,132,000

(b) From Equation in Section B.1:

Assets = liabilities + net worth
$1,132,000 = 603,000 + 529,000

B.2 Determine the decrease or increase from July 1 to July 31 in materials inventory.

25,000 - 46,000 = $-21,000 (decrease)

B.3 (a)

Non-Stop
Income Statement
Month Ended July 31, 20XX

Total revenue		$500,000
Expenses		
Cost of goods sold	$141,000	
Insurance	20,000	
Selling	62,000	
Rent and lease	40,000	
Salaries	110,000	
Other	62,000	
Total expenses		435,000
Income before taxes		65,000
Income taxes		20,000
Net profit for month		**$45,000**

It is necessary to construct the Cost of Goods Sold statement to get the $141,000 amount.

Non-Stop
Statement of Cost of Goods Sold
Month Ended July 31, 20XX

Materials			
	Inventory, July 1	$46,000	
	Purchases	20,000	
	Total	66,000	
	Less: Inventory, July 31	25,000	
	Cost of materials		$41,000
Direct Labor			50,000
Prime Cost			91,000
Overhead charges (Indirect costs)			75,000
Factory Cost			166,000
Less: Increase in finished goods inventory			25,000
Cost of Goods Sold			$141,000

(b) From Section B.2

Revenues − Expenses = Profit
500,000 − 435,000 = $65,000 (before taxes)
500,000 − 455,000 = $45,000 (after taxes)

(c) Percent in after-tax income $= \dfrac{45,000}{500,000}$ (100%)

$= 9\%$

B.4 (a) Current liabilities = accounts payable + dividends payable
= 35,000 + 8,000 = $43,000

Current ratio $= \dfrac{\text{Current assets}}{\text{Current liabilities}}$

$= \dfrac{77,000}{43,000} = 1.79$

Acid-test ratio $= \dfrac{\text{Quick assets}}{\text{Current liabilities}}$

$= \dfrac{\text{Current Assets} - \text{Inventories}}{\text{Current liabilities}}$

$= \dfrac{77,000 - 31,000}{43,000} = 1.07$

$$\text{Equity ratio} = \frac{\text{Total net worth}}{\text{Total assets}}$$

$$= \frac{529,000}{1,132,000} = 0.467$$

(b) This value may be computed in two ways:

1) $\text{Percent tied-up} = \frac{\text{Inventories}}{\text{Current liabilities}} (100\%)$

$$= \frac{31,000}{43,000} (100\%) = 72\%$$

2) (current ratio - acid-test ratio) 100%
$= (1.79 - 1.07) \, 100\% = 72\%$

B.5 (a) $\text{Net sales to inventory} = \frac{\text{Net revenue from sales}}{\text{Average inventory}}$

$$= \frac{500,000}{31,000} = 16.13$$

This means the average inventory value of $31,000 was sold (it turned) 16.13 times during the month of July.

(2) Determine the return-on-sales ratio.

$$\frac{\text{Net profit}}{\text{Net sales}} (100\%) = \frac{45,000}{500,000} (100\%) = 9\%$$

Note: This is, effectively, the same as the question in B.3(c).

(c) The profitability index is the value of the return-on-assets ratio.

$$\frac{\text{Net profit}}{\text{Total assets}} (100\%) = \frac{45,000}{1,132,000} (100\%) - 100\%$$

$$= 3.98\%$$

The median from Example B.1 is 1.8%, so Non-Stop is doing well for an airline.